MAN AND ANIMAL

ANIMAL HAND AND HUMAN HAND

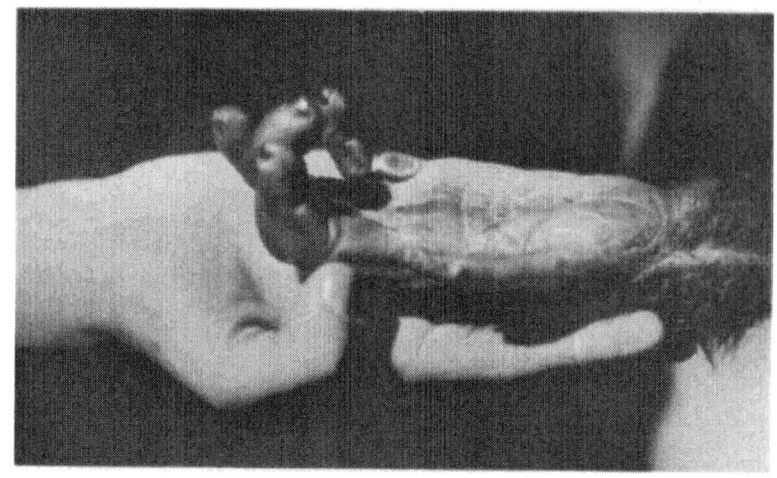

HAND OF CHIMPANZEE LYING IN A HUMAN HAND—A CONTRAST

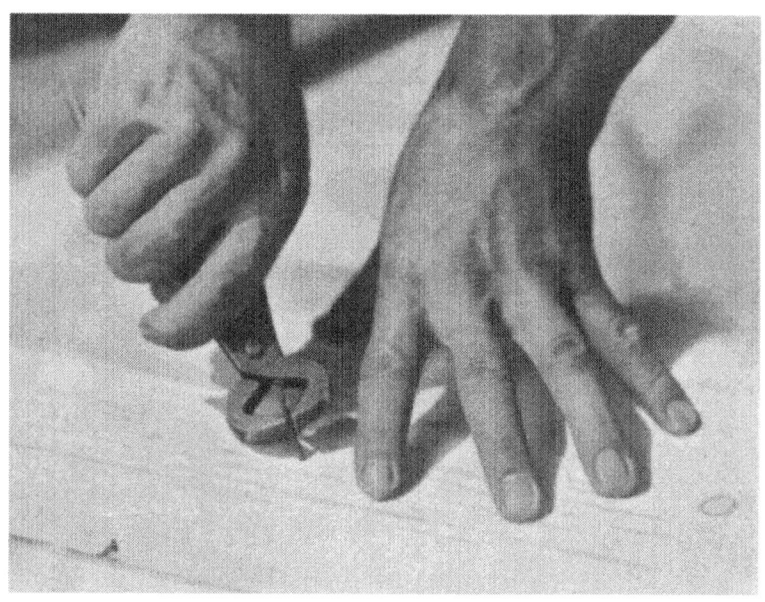

HUMAN HANDS WORKING WITH A TOOL

(Reproduced by permission of Carl Schütze, Hamburg)

MAN AND ANIMAL
THEIR ESSENTIAL DIFFERENCE

by

HERMANN POPPELBAUM, Ph.D.

Edited by the Natural Science Section of the Free High School for Spiritual Science, at the Goetheanum in Dornach (near Basel), Switzerland.

TRANSLATED FROM THE GERMAN

RUDOLF STEINER PRESS

Rudolf Steiner Press
Hillside House, The Square
Forest Row, RH18 5ES

E-mail: office@rudolfsteinerpress.com

www.rudolfsteinerpress.com

First published in English translation in 1931
Second English edition, based on the sixth German edition and edited by
Owen Barfield, 1956
Reprinted 2014

First published in 1931 under the title *Mensch und Tier* by Verlag am
Goetheanum, Dornach, Switzerland

© Rudolf Steiner Press 2014

All rights reserved. Apart from any fair dealing for the purpose of private
study, research, criticism or review, as permitted under the Copyright,
Designs and Patents Act, 1988, no part of this publication may be
reproduced, stored in a retrieval system, or transmitted in any form or by
any means, electronic, electrical, chemical, mechanical, optical,
photocopying, recording or otherwise, without the prior written
permission of the copyright owner. Inquiries should be addressed to the
Publishers

A catalogue record for this book is available from the British Library

ISBN 978 1 85584 407 0

Cover by Morgan Creative featuring an illustration © Velazquez
Typeset in Great Britain
Printed and bound by 4Edge Ltd., Essex

EDITOR'S PREFACE

WHAT SCIENTIFIC research stands in most immediate need of to-day is, not so much an increase in the number of individual facts—the heap is already piled so high that we cannot see over it—as a clear decision one way or another concerning a few of the cardinal questions of Natural Science. First and foremost among these is the historical and evolutionary relation between man and animal, a question which, either because of Darwin or in spite of him, still remains unsettled. Yet a decision on this one point must settle conclusively one way or another a whole host of theories and interpretations which have found their way into the outlook of large sections of mankind to-day and are influencing their daily life. Huge edifices of thought have been erected on foundations whose ability to bear them is questionable to a degree. Dr. Hermann Poppelbaum has made investigations of decisive importance into these fundamental problems and gives us the result in the present work. The Natural Science Section of the Free High School for Spiritual Science at the Goetheanum, Dornach, especially welcomes the publication of this book, which is not merely one small stone in a great temple of learning but also a true foundation-stone for a new and securely based outlook on the world. The possibilities of building farther on this contribution are infinite; they deserve the bold co-operation of every creative Natural Scientist.

For the Natural Science Section,
GUENTHER WACHSMUTH, Ph.D.

Dornach, June, 1928.

THE AUTHOR'S INTRODUCTION

THE FIVE ASPECTS here taken are the outcome of a search which began in the author's very early youth. Even then his thoughts were busied, in a boyish way, with the enigma of animal existence; but no ordinary curiosity drove him on. As the years passed, he saw more clearly that the difference between man and animal was not merely theory to be handled by the analytical intelligence; but instead, that here was a problem of profound significance for humanity. The secret of the two kingdoms—so closely related to one another, yet separated by so profound an abyss—drew him ever and again under its spell.

With passionate interest the author followed the struggle and the triumph of the theory of descent. He saw with what great success the champions of the animal descent of man brought forward evidences of transitional stages of bodily development. He watched how, when the physical difference became unreal, hard-pressed objectors were forced to take shelter behind doubtful definitions of a spiritual difference between the two kingdoms. And because in the first ten years of the new century the noise of the combatants had died down and silence followed—it seemed to him as if the final word had been spoken.

Then nearly ten years ago another phase began for him with his fateful meeting with Rudolf Steiner's Anthroposophy. A whole world of new questions opened out concerning things about which he thought he had satisfied himself long ago. With astonishment he recognised that here was the basis of a new conception of the world, laid down as far back as the eighties of the last century; and that to-day, when the old ideas were crumbling, was no time for mere passive acceptance. For decades Rudolf Steiner had been at work on this new structure.

Amidst the endless wealth of new knowledge—unlocked for the writer through this meeting—was an entirely new, bold delineation of the essential relationship of man and animal. Man, no longer the accidental finis to the animal species, but the centre of gravity in the kingdoms of nature. . . . Man, not the final product of evolution, but from primordial times its hidden source. . . . The human soul, no longer a further and more highly developed animal soul, but the harmonious and uncorrupted archetype in which the animal soul is contained as a part. . . . The human spirit no longer the reflex of the nervous system, but the microcosmic reflection of the Divine. . . . The whole man the key to the physical and spiritual universe. . . . Man's moral entelechy, evolving in his earthly life, the seed of a cosmic future. . . . Such a delineation as this could not but take powerful hold of the writer, making him return with all his strength to the old " question of questions " concerning man's origin.

Here was a man able to put questions, and to give the answers in such wise that all merely theoretical knowledge, all satisfied dependence on traditional thinking was made contemptible. A man able to put weariness to flight, and to open new sources of inspiration: who convinced every man—who would listen—that at this point his most personal, his greatest treasure, his very being as man was challenged. Here were no mere tedious, idly propounded problems, but questions which arose from the innermost need of the heart. And his replies were not of the adroit kind, framed chiefly to intercept further questioning. Man's pursuit of truth was wrested from the enervating ripple of mere argumentative debate, and was set in the full flow of devoted research, of the mighty stream of new knowledge; wherein man became conscious of himself and could speak with the whole world and with every being. The answers given were not attempts to silence him, but such as sought to fill him anew with life. In the immense germinating power of this new life, the learner began to see that where this great source of fruitfulness was, there Truth itself must be.

It was this wonderful power of putting new questions and of giving the answers to them which was Rudolf Steiner's gift to mankind. And thus it is that the contents of the accompanying work are unthinkable without that gift. In it both the new turn of the questions and the characteristic new answers are due entirely to the great teacher. In this respect it is hoped that the book may play its part in the awakening call which rang out from Rudolf Steiner's work.

There was, however, another incentive to publish. Quite recently some distinguished scientists have published certain theories which herald a reversal of the theory of descent. The palæontologist E. Dacqué, in his book *Urwelt, Sage und Menschheit* (*The Prehistoric World, Legend and Mankind*, 1924) ascribes to the human race a very great geological age, an age so great as to make it impossible that present animal stocks were the progenitors of man. The Dutch anatomist L. Bolk in a lecture at Freiburg (1926), demonstrated that the human body is not a further development of animal form, but on the contrary, that it has remained backward in relation to animal evolution. The pathologist M. Westenhöfer astonished the Anthropological Congress at Salzburg in September, 1926, with his thesis that the higher mammals have evolved from man, instead of the reverse process. If there were any who expected that statements like these would make a great commotion—as twenty years ago they certainly would have done—they were disappointed. Jaded indifference and superficial scepticism has hold of the public. A few sensational Press notices were answered by superficial contradictions, and public attention soon turned to other things.

Nevertheless those new conceptions opened up the way to a knowledge which Rudolf Steiner had indicated decades before. But they receive their full value only when included (and with this they are stripped of paradox) in the whole picture of spiritual evolution.

Such indications of public apathy prove that to-day quite other forces are needed than sufficed formerly to secure attention for the problem of the true being of man and animal. A new hypothesis, no matter how forceful and bold, if thrown singly into the discussion, is no longer any good. The day of the specialised problems is over. The whole being of man is in question!

Thus the call for an anthroposophical presentation of the problem was redoubled. In very truth our sight must be trained to a universal range; only so can it include the inner being of man; only then shall we be shaken out of our present inertia, and awakened to a sense of the gravity of the question at issue.

The present book is a most imperfect attempt to survey the boundless waste of problems that beat upon us; and to focus them from five different aspects. The reader may forgive its fragmentary character when he realises that all five converge upon an invisible centre. In this centre there lies the real being of Man. That the sight is directed to this centre constitutes the value of the book. All observation must remain superficial unless it advances to the stage where contemplation is transmuted into moral experience.

But at this point the reader must be left at liberty. This book will offer him no theory; but it will teach him to see for himself. He must decide of his own free will. In this respect the author seeks to make his own that quality which the great teacher possessed in such a remarkable degree:

Confidence in the germinating power of Truth.

HERMANN POPPELBAUM, Ph.D.

Frankfurt a.M., January, 1928.

INTRODUCTION TO THE REVISED (SIXTH) EDITION

THIS BOOK, which sets forth Rudolf Steiner's contribution to spiritual-scientific research on the subject of the descent of man, has now been before the public for twenty-eight years. It may be said to have exerted some influence—and not only in quarters where the influence is admitted. Frequently indeed the basic ideas in it have been simply appropriated without acknowledgment.

The author has nothing to add to the 1928 Introduction, except to say that he has been concerned, as before, to bring out the congruity between the findings of spiritual science and the *actual results* of morphology, palæontology, embryology, psychology and epistemology. In Chapter 2 he has paid particular attention to important advances in our knowledge of human and pre-human fossils; and it proved necessary to rewrite altogether a part of this Chapter ("Descent"), replacing illustrations and diagrams by more up-to-date ones. Recent discoveries have confirmed in a gratifying way the fundamental soundness of the anthroposophical approach.

We are well aware that the radical use which the author has thought fit to make of the new and extended discipline of knowledge known as Anthroposophy is an obstacle to the acceptance of this book in scientific circles. That is, however, a stumbling-block which, in the interests of truth, he could not bring himself to remove.

HERMANN POPPELBAUM, Ph.D.

Goetheanum, Dornach, Switzerland, May, 1956.

CONTENTS

	PAGE
EDITOR'S PREFACE	v
AUTHOR'S INTRODUCTION	vii
AUTHOR'S INTRODUCTION TO THE SIXTH EDITION	xi
PART I FORM AND SHAPE	1
II DESCENT	27
III SOUL	85
IV EXPERIENCE	109
V DESTINY	136
NOTES	153
LIST OF AUTHORS REFERRED TO IN TEXT	163

ILLUSTRATIONS AND DIAGRAMS

Hands of Man and Chimpanzee . . . *Frontispiece*

		PAGE
1.	Young and Adult Chimpanzee	8
2.	Skeletons of Penguin, Gorilla, Man and Kangaroo	20
3.	Haeckel's Tree of Descent of the Vertebrates	32
4.	Table of Descent according to Gregory	34
5.	The Tree of Descent according to Present-Day Research	37
6.	Changes in the Table of Descent of the Crab, 1873–1912	42
7.	Lemurian, Atlantean and Post-Atlantean	62
8.	The Spiritual and Physical Aspects of Evolution	63
9.	The Reflection of the Four Earth Epochs in the Life of the Human Embryo	72
10.	Embryonal Stages of the Crocodile, Hen, Ape and Man compared, illustrating the Law of Retardation	77
11.	The Degeneration of the Skull of Orang and Gorilla compared with Embryonal and Adult Human Skulls	79
12.	Table of the Rhythm of Development in Man and Ape	82

13.	Heads of Human and Gibbon Embryos . . .	83
14.	Stages of Incarnation of the Four Kingdoms of Nature (from Rudolf Steiner)	103
15.	Diagram of the Inwardness peculiar to "Creatures of Instinct"	116
16.	Sketch showing the difference of Consciousness in Man and in Higher Animals	135

PART I

FORM AND SHAPE

Every creature appears before man with bended knee.
—K. E. v. BAER.

ANYONE WHO TRIES to compare the bodies of man and animal must—to an unusually high degree—free himself from prejudice. No preconceived ideas should be allowed to disturb his tranquil examination of the facts. His eye must dispassionately search out the differences of form, where they are actually to be found; those which are accepted by traditional opinion must not be magnified, nor those suppressed which are inconvenient. Neither to coincidences nor to discrepancies must more importance be conceded than actually belongs to them.

Such liberty from prejudice was lacking in the foremost investigators of the second half of the nineteenth century, but who would reproach them for it? It would be ingratitude, for to their passionate interest both anthropology and zoology owe their most significant discoveries. Yet to-day any investigation of the physical differences between man and animal must begin by deliberately freeing itself from all unproven beliefs left in the trail of the battle over the descent of man. Questions as to which forms are earlier or later, which are more primitive or more perfect must be postponed altogether and our conceptions developed with a calculated restraint purely from observation of the forms themselves.

Investigators who strove to do this were not lacking even in the last century, but in the heated discussions of the time their voices went unheeded. They were accounted eccentrics, whose views were beside the point, and unlikely to produce any new knowledge.

Yet it is precisely to such unorthodox ideas, to trains of thought which run counter to settled opinion, to conceptions differing widely from the accepted theories, that we owe all progress. The man who will accept only what he immediately comprehends remains spiritually at a standstill; it is the person who seizes with daring the thought that first repels, who strikes from it in the end the spark of some surprising revelation.

HAND

The reader should test this for himself. Let him consider that much lauded miracle, the human hand, and set it in his mind's eye beside the various forms which take its place among the animals: the lion's paw, the horse's pastern, the digging foot of the mole, the sloth's climbing foot, the bird's wing, the fish's fin. Let him give himself up patiently to the consideration of this series of well-known forms, passing from one to the other, and as often reverting to the human hand. Let him try to make clear to himself what all these various limbs, considered as physical structures, really *are*, and provided that he leaves all his long-cherished conceptions of perfect and imperfect on one side, he will be obliged to admit that the human hand is of all these forms the *least* well equipped physically for its tasks. That it is a wonderfully perfect tool one has heard hundreds of times, yet in physical aptitude it is far behind the corresponding extremity in an animal. Certainly the hand is incomparably better articulated, of more universal application, more deft, more mobile than any claw, fin or wing, nevertheless it is not an *instrument* in the sense in which these animals limbs are. It lacks something of that perfection which they all possess. Leaving out of account its infinite latent possibilities, then, considered purely as a tool, formed for a particular task—the hand is the most imperfect of all: and as long as we consider solely the fundamental physical form, it is the form of the hand alone which can be taken as the point of departure for all others. None of the others is fitted for this role. Not the fin, which is only an oar; not the horse's hoof: its sphere of utility is much too limited; not the wing, for it serves only in the air. Only the hand includes all these in itself—

and that not because it has been perfected beyond all the others, but, on the contrary, because it has remained behind all the others. Physically considered the hand is undoubtedly a more primitive form than any other extremity and if we are to regard all these extremities as having developed from one primal form, then we shall be obliged willy nilly to imagine a handlike form as the starting-point. It avoids any excessive finality, any unbridled metamorphosis which must necessarily limit its adaptability. The secret of its versatility lies in its arrested physical development. It is able to become the tool of the human spirit just because of its physical shortcomings.

This peculiar relationship becomes obvious (and the reader should confirm this second step with careful attention) when we recollect all the cunning tools devised by the mind of man to replace numerous shortcomings of his hand. A man must lengthen his hand, so to speak, by means of the oar to make it into a fin (an oar is a fish's fin made of wood); he must put on boots in order to run well (the horse's hoof gave the original design for the boot; which, by the way, is often braced with iron). He straps climbing irons on his legs when he needs to climb (the climbing iron is the sickle-shaped claw of the sloth, copied in steel); he needs a breadth of surface, so as to be able to fly (the plane of the aeroplane is the bird's wing made rigid). Only when the extremities of animals are examined in this way do we begin to comprehend what it means to equip a body with instruments; we discover how strikingly the animal organs resemble those dead mechanical tools invented by man, whose perfection is the greater the more limited their purpose. The hand alone preserves its dexterity through an infinite variety of functions because in it the specialised tool is less in evidence than in any animal limb. A distinctively human attribute, the latent creative skill of the hand, is rendered possible by the retrogression, so to speak, of its physical form, a conclusion which—at first repugnant to us—can nevertheless lead us deep into the heart of the matter.

We are all too much accustomed to thinking of the lower as the undeveloped and the higher as the developed; otherwise it would not be nearly so difficult as it is to see that the more primitive form may conceal within itself greater possibilities and in this respect actually be the *higher* of the two. If we succeed in forming some sort of a mental picture of the original form of an extremity, we can then imagine at will this or that animal limb, evolving from it, sacrificing its multiform adaptability, as it goes on, in favour of some single specialised development. And in this way we realise how the human hand stands nearest to this fundamental type, and betrays least development beyond it. We discover that the fundamental type of an extremity can manifest itself in the most varied forms, every one of which can be seen with the "inward eye" to be evolving out of it. In the case of the human hand, we divine its proximity to this invisible archetype. Structurally it stands nearest to the centre, with the others at the periphery. Any system which represented this relationship fairly must show the human hand in the midst, surrounded on all sides by animal limbs, each one of which has diverged from the centre in a different direction. We need such a representation to replace the forced and erroneous genealogies by means of which people have tried in vain to demonstrate the evolution of the hand from the anterior extremities of ancestors of an animal type.

If anyone suspects something artificial or forced in this, he need only consult that earliest natural record in which the organism can be seen developing before our very eyes—embryology. The facts of embryology accord in a striking way with what has been said above. If the embryos of a man, a dog, a bat, or a dove are observed in their early stages, it can at once be seen how hand-like the anterior limb is at first, the future form only gradually developing from this hand-like origin. The dog's paw and the dove's wing diverge with rapid strides from their resemblance to the hand, to become at last mere tools for running or for flight. It is only in the human embryo that the hand preserves its essential proportions, retaining them so definitely that, when fully grown, it still reminds us forcibly of its original form. We observe, therefore, that a certain stopping short is the characteristic mark of the human being, while in the animal the corresponding limbs tend towards a definitive metamorphosis. With the embryo hand as centre, they diverge in all directions, all the time assuming more and more distinctly the character of perfected tools, until at last, at the end point of their own special line of development, they come to a standstill. The whole forms a natural document, proving the accuracy of that radial system of classification which was arrived at in the first place on the basis of comparison.

The " record of the rocks," or (to use Haeckel's terminology again) the " palæontological record," demonstrates the same point. Every textbook contains the now famous " Genealogy of the Horse," showing the gradual development of the legs of the modern animal as recorded by Nature herself in the various strata of the tertiary period. The contemporary horse has only the third member of the radial five-toed system to run with; it runs, one might say, on an abnormally strengthened middle finger. The remaining radii, as the ancestral chain shows, have degenerated, though only very gradually. The primal form from which we start (Eohippus) has the five radial bones still fully developed; but without one of them being lengthened into a special running limb; all are quite short, or, in other words, the extremity as a whole is still very similar to the human hand; while in its further development to the present stage it has left this similarity behind. The fact that this is the true lesson of the favourite textbook illustration is (unfortunately) generally overlooked (see Note 1).

All three of Haeckel's sources, the findings of comparative anatomy, of embryology and of palæontology, bear witness to the fact that the human hand has remained nearest to the primitive organ, while it is the animal extremities which are the more developed. The former has retained the universality which the latter have given up. Once this objectionable notion has been grasped—that an advance in evolution may have to be paid for by some form of bodily sacrifice, it is not difficult to see that this evolutionary principle operates on a far greater scale than one at first supposed. Whole areas of our human physical structure, as distinct from that of the higher animals, begin to emerge in a new light.

HEAD

This is true even of that proudest possession of the human body, the head. It is precisely the human head of which people are always saying that it is "developed" far beyond the animal stage. But is this really the case? Are we not deceived by our preconceived notion that man is, physically, the culminating point?

If we compare the structure of the human head with that of any animal, what strikes us most forcibly is the tremendous difference in the face. With the human being, there is a harmony between the brow-and-eyes area on the one hand and that of nose, mouth and chin on the other. Whereas in the animal all this is displaced. The head, without any well-defined brow, is drawn out into a nose or snout. Nose and jaw attract to themselves a disproportionate share of the forces that go to form the head, leaving no possibility for the development of a proper forehead above or chin below. Thus we cannot properly speak of an animal *countenance*. Examine the long profile of the horse with its relatively small muzzle or the shortened profile of the cat with its wide jaws. They are characteristic animal heads. Again, if the head of an ape is compared with the human head it is at once manifest how the former diverges from the latter in a definitely bestial direction. There is here the same over-development (compared with man) of the nasal area as in all the mammals; the brow is scarcely, the chin not at all, developed; and the nose instead of dividing the face is drawn down to the mouth and stuck upon it without any real prominence. Even amongst the man-like apes, in the case of adults, the proportions are not much better. Not so much a face, one might almost say, as a glorified muzzle.

TEETH

The inside of the mouth discloses an excessive development of teeth and jaw. Instead of the even row of human teeth, we get a true animal jowl with teeth of the most varied shapes. Apes in particular have all of them a fearful weapon in their jaws; the corner teeth project in sharp crooked cones, and, especially in the male, are barely hidden by the full lips.

Here again, if we enquire which of these two types of jaw is the earlier—only one answer is possible: the human; the animal face could never be shortened to resemble a human one; its powerful teeth could never dwindle; nor the coarse lips be refined to human proportions. No, the opposite is the truth. The animal head is a kind of distortion of the human original. We can imagine the heads of animals as having evolved from a human head by the exuberant over-growth of some one particular region. It is unquestionable that the animal head is the most developed and the human head the form that has remained behind. The human head lacks something with which the animal head has been richly endowed.

If we ask what it lacks, we find that—just as with the hand—it is aptness for use as a tool. Man's head is entirely removed from active participation in its surroundings; it holds aloof from all direct intervention; it reposes in freedom on its lofty height, wholly dedicated to looking and listening; during even the most violent activity of the body the head can look on comparatively undisturbed; it lets itself be fed like a brooding bird. An animal's head, on the contrary, is continually being claimed for physical uses. Even where, as with the bird, it is held high, it is used as an extremity, both for carrying food and for tearing it to bits; for work on the nest; for cleaning its feathers; for defence from attack; it is used, in short, as a tool! First hand necessities and requirements have given the animal's head its form. The whale's head has to part the waters; the mole's to tunnel the earth; the bird's to cleave the air; the beaver's head is a chisel; the parrot's a pair of tongs; the wood-pecker's a hammer and an axe.

The perfecting of the animal head as a tool for physical operation on its environment robs it of the universality that distinguishes the human head: but what it sacrifices in universality it gains in physical efficiency. Of this we can find convincing proof by comparing the skulls of man and ape. What a wealth of protuberances and accretions at the sides, in front, above, below! The rims of the eye-sockets are thickened rampart-wise into a kind of bony pair of spectacles and separated by great swollen eyebrows from the receding forehead. On the crown of the head rises a bony ridge which joins at the back the ridges running from both temples, and here, at the place where in man the cranium and occiput are rounded off like a dome, in a wonderfully smooth curve, just here, we find in the ape a sharp edge! These extensions and additions to the skull provide leverage for certain powerful muscles of the head and neck, and combined with the dominant jaw and teeth they give the ape's skull its bestial expression.

This character, however, is only developed within the course of each individual life from a skull whose form is at first clearly reminiscent of human proportions. The new-born ape still has a beautifully domed skull, and a face properly set back (Figure 1).

Figure 1.

Young Chimpanzee.   Adult Chimpanzee.
(Reproduced here with the kind permission of the publisher: Julius Springer, Berlin. Taken from J. Naef's work in the 20–21st number, fourteenth year of the magazine *Die Naturwissenschaften*.)

There are few sights more estranging than this change from the well-defined rounding of the back of the head and brow, and from the delicately formed nose and mouth to the appallingly ugly, almost canine profile of the mature male chimpanzee. And this unrecognisable distortion takes its course in four to five years; i.e. in a space of time in which the face of a little human child has scarcely altered. While the child's face retains for a long time something like its original form, in the ape there takes place such a loss of all human characteristics as well nigh moves us to pity. "From now on," writes the Swiss, L. Rutimeyer, 1868, "the most awful brutality grins between its protruding teeth from this countenance which only a few years before had such a pensive expression." Involuntarily the question escapes us "What has become of you?"

The human head does not accompany it in this striking deterioration, but attains to its own true significance by keeping its pristine shape. It remains behind physically, and in this way serves progress.

BRAIN

And the brain, the paramount foundation for this progress—has this not evolved beyond the animal's? It has—but not by a further development of the animal form; on the contrary, because it continues to maintain more closely than that of any animal the relative proportions of the embryo during the last few months before birth. Here again, the basis for an advance is the retention of the original inter-relationship of the various parts. The Amsterdam anatomist, Bolk, has illustrated this very clearly. The axis of every vertebrate embryo is considerably curved at the cranial end—rolled back on itself like the handle of a walking-stick. This curve, which Bolk has examined and named in detail, in man maintains its characteristic form throughout life: speaking broadly, this is the reason why, in a full grown man, the nose still points downwards in a direction almost parallel to the axis of the body. On the other hand, among mammals, during the fœtal period, a complete straightening out of the curves takes place. This can be very clearly seen when the profiles of man and dog are compared. Bolk says of this, "Indeed no part of the human body is stamped so characteristically with the persistence of the fœtal condition as the head" (1926, p. 31). Thus the head and brain of the higher animal have, to begin with, human proportions, but cannot retain them.

By the straightening out of the anterior curves of the body's axis, the animal's head takes a new direction, along which it is then obliged to continue growing, just as the brain itself is held fast within the hardening skull-case. In man, on the contrary, the skull-sutures coalesce only during the second decade: this allows the human brain to increase in mass during this period without having to relinquish its fœtal proportions. The truth is: *The brain can attain human development precisely because it is able NOT to go on evolving in the animal direction, but to continue elaborating the periphery and the various centres without at the same time losing the fœtal proportions.* The human brain is not a more highly evolved animal brain: but, on the contrary, one that has remained plastic for a longer time and kept its original proportions. (In man the foramen magnum maintains its embryonic position, but in the animal head there is a shifting in a backward direction: Bolk. See Note 2.)

It was Karl Ernst v. Baer—called by Haeckel the father of Embryology—who once gave a humorous sketch of the sort of thing we might look for if birds had to compose an embryology of mammals. Usually the latter order is placed at the head of the vertebrates. Karl Ernst v. Baer thinks the birds might well produce some excellent reasons against this theory, as in the following extract from a textbook on *Embryology for the Use of Birds*:

"These quadrupeds and bipeds remind us in many ways of our own embryo stage: their skull bones are separated; like ourselves during the first five or six days of incubation, they have no beak; their extremities are all pretty much alike, as ours are during about the same period. Their bones are solid, containing, like our own in early youth, no air. There is not a single feather on their bodies, which grow only thin stems, so that before we have emerged from the nest we have already passed the extreme limit of their development. They lack the most rudimentary crop, and their anterior and posterior stomachs are practically fused together into a single sac (both purely temporary conditions in our case). Their nails betray in many cases that awkward flatness which is characteristic of ours before we come out of the egg, and as to the flight faculty—it is confined entirely to the Bat, apparently their most highly evolved genus. And these mammals, incapable of finding their own food for months after birth and permanently incapable of raising themselves from the ground—these mammals will have it that they are more highly organised than we are!" (1828, p. 203).

These delightful and not less pregnant words could be applied *mutatis mutandis* to the development of the human embryo as described by the apes. With a little imagination the reader may, from the facts here given about the hand, head, jaw and brain of man, work out for himself a fictitious " Embryology of Man, for the Use of the Monkey Race," in which the unassailable truth that man owes his advance to the fact of his remaining physically *behind* the ape can be used to impugn his claim to superiority. It would be easy for any mammal to prove that it is farther advanced in its mother's womb than the human being at the end of his life.

HAIR

The literature of comparative anatomy and embryology will supply an abundance of facts to prove that the principle of arrested development apparent in the form of the human body, and illustrated above in respect to hand, head and brain, applies also without exception to every human characteristic. Significant in this respect are L. Bolk's findings as to the hairy covering in man and ape (1926, p. 26). It is a well-known fact that in the human being before weaning, hair grows solely on the head, though most animals are born with a hairy covering: later on and particularly at the age of puberty, a slight hairy growth may appear as a later development on some parts of the human body; but even this is sometimes lacking, as amongst Mongolians. That there is a latent tendency to a complete covering of hair is shown in those cases where through inner disturbances the normal restraining forces are absent, and so-called hairy-men (Hypertrichi) appear. Thus, in the animal, that further development has taken its course, which is restrained in man. Bolk studied the development of the hairy covering of the apes, and found that all the lower apes fall into line with the other animals, but that among the man-like apes hair is at first developed only on the head, the rest of the body being quite naked, until it appears on the body during the first two months after birth. Here again we see clearly that man comes to a halt at a stage which the animal passes through, and that man maintains for his whole life a condition corresponding to that of the fœtus, or indeed to that of the new-born man-ape. It is remarkably significant that the new-born chimpanzee or gorilla should be as naked as a human child and should have fairly long hair on the head alone (Bolk, 1926, p. 27). Profoundly impressed by these discoveries, Bolk expressed this difference between man and animal in a radical way, carrying to a conclusion which has already excited considerable opposition, the view that man develops conservatively and animals precipitately. From the physical point of view, he said, " man is really a primate fœtus which has attained to puberty as such." This is admittedly rather a sensational way of putting it, but it contains a kernel of truth well qualified to arouse an observer caught in the grip of conventional theories and to open his eyes to a paradox which is at the same time a genuine fact. And that is urgently needed.

It is significant that it should be precisely about the hairy covering of the human embryo that the most misleading ideas have been broadcast. We find in so many popular textbooks dealing with the theory of descent the statement that the human embryo at a certain time before birth has a furry skin like an animal's, and thus that it visibly passes through an animal stage. Against this, one can convince oneself by studying the careful researches of Hans Friedenthal that in the fœtus of the third or fourth month there is in fact no furry hide at all, but a covering of down, a coat of extraordinarily fine little hairs, the relics of which are kept for life. This downy covering moreover lacks the scent or feeler hairs possessed by all animals, including the Anthropoids, and which are surrounded at their roots by a blood sinus. The animal's ultimate hide is *additional* to this coat of downy hairs, and appears indeed shortly after birth or just before. It is really misleading to speak of the human fœtus being hairy like an animal, for the particular thing which in animals becomes fur, remains among men restricted to slight indications and literally *un*developed. Friedenthal, who when he was engaged in these studies (Note 3) knew nothing of the wide scope of the phenomenon of arrested development, nevertheless writes with a certain instinctive feeling: " Death sets a term to the life of man, just as the permanent covering of hair is *beginning* to spread over the entire surface of the body." ... " In his hair, as in many other things, the retention by man of his youthful characteristics beyond the age of sexual function and right up to death, shows that the term or goal towards which he is tending is not reached in the course of an individual life..." (*Man and Ape*, pp. 20 and 37). This supposititious " goal " is of course the animalism of the human form: though " goal " is merely a misnomer for the very thing that can never happen. This is another example of the deep rootedness of this idea of the " primitive " structure of the animal. It will be a long time before this idea is got rid of.[1]

KIDNEY, SPLEEN AND CÆCUM

Not long ago M. Westenhöfer made a considerable stir at the Anthropological Congress in Salzburg (1926) when he brought forward his thesis that unlike other higher mammals, man has *retained* in his inner organisation (kidney, spleen and cæcum or blind-gut) certain distinct features by which he is closely related to the primitive aquatic mammals. He names these primitive structures Progonisma. The apes in particular have developed these organs far beyond the human stage—and cannot therefore

[1] Shortly after the publication of the first edition of this book, O. H. Schindewolf (1928) published his essay, in which he suggests that humanity was attained by way of proterogenesis. He means by this that to the characteristics first noted among the invertebrates, new ones were added, and were passed on by heredity. The new acquisitions appear at first only in the young of the succeeding generations, to disappear again in maturity; later, however, they are retained throughout maturity. Schindewolf sees the resemblance of young apes to men and their subsequent relapse into the bestial as a case of proterogenesis.

The parallel is certainly interesting, and yet Schindewolf remains fettered by the current doctrine of heredity, when he speaks of " feature-complex man " (*Merkmalskomplex Mensch*) (see Review—" A Notable Idea concerning the Birth of Man "—in the weekly periodical *Das Goetheanum*, 1930).

be classified as more primitive. Obviously the researches of the Berlin pathologist provide one more illustration of this fact that the form of the higher animal is a more developed, the human form a more backward one.

CASES OF ATAVISM

We may further recollect the large number of freaks in the human form which the defenders of the animal descent of man were fond of bringing forward in proof of their claim; for instance the famous " pointed ears " which the Darwinian School accepted as a throw-back to animal-like ancestors with ears of this kind. When the occasional enlargement of the " crypts of Morgagni " in the human larynx was discovered it was regarded as a reminiscence of the howling cavities (Brüllsäcke) which are still to be found fully developed in Anthropoids (see page 21). Furthermore when Meirowsky (1920) actually proved that the birthmarks in man are related in their distribution to the markings on the skins of mammals, there was great exultation in some quarters. And yet a little reflection should prove to us that, while there is a tendency to these in the human body, ordinarily they are not developed, but restrained; this is exactly what was emphasised in the foregoing pages: man has in his body a tendency to the animal forms, but it is held in check and does not come to expression. Thus, the anomalies in question are in no sense throw-backs; but rather formations which have overshot in the direction of the animal the check normally exerted by the human form. Instead of pointing to the ancestors from which the human form has sprung, they show what this form can become when the retardation of this development is not maintained (Note 4).

GENERAL SURVEY

Many additional details might be given here; we content ourselves with commending to the reader the authors mentioned in Note 4: Virchow, Lucae, Eimer, Selenka, Alsberg, Boule, Schwalbe, Stratz and Bolk. More useful than the most complete enumeration of such details is the attempt to make a survey of the phenomena of retardation, and to provide a complete picture of their effects on the human body. This attempt can be made only on the basis of Rudolf Steiner's epoch-making indication as to the architectural plan underlying the human form. And here our considerations must at once overstep the boundaries customary in comparative anatomy and we must regard the human form in connection with its earthly and cosmic environment. For it is only in this way that we shall find the inner connection between the "primitiveness" of the human physical organisation and its significance for the whole nature of man.

As early as 1775, Goethe pointed out, in the following incomparable words from his *Physiognomical Fragments*, the difference between the erect human form and the earthward-bowed body of the animal. "In truth the whole form stands before us like the central pillar that supports a vaulted roof—a dome which is to mirror heaven. How like the heavens is the dome of our skull above us, made to hold the pure image of the circling spheres. . . . And how perfect a contrast is the structure of the beasts! The head merely depends from the spinal column. Subsisting, as it does, principally for the purposes of sniffing, chewing and devouring, the muzzle and the throat are its most excellent parts. The muscles, flat taut, and covered over with a crude harsh skin, are incapable of any refinement of expression."

Rudolf Steiner described in a comprehensive way, drawing from spiritual sources to which he had access, the polarity of the upper and lower systems in man's organisation, and the middle system which unites these opposites (Note 5).

The upper pole of the human body is most strongly withdrawn and estranged from the earthly forces of gravity; the head therefore rests on the neck, free, even when man's body is in movement. The lower pole, on the contrary, is completely at the mercy of the earth forces: the legs which carry man are dominated entirely by the architectural principle of the pillar, a fact which appears quite clearly in the skeleton; the form of the human foot is a witness of this inter-relationship with the earth's field of gravity. The head, on the contrary, has become the image of quite different forces, which instead of raying out from earth's centre of gravity, pour down from the infinite periphery, seeking to round off the skull to a sphere (Note 6). Between these two poles the breast system is the mediator, neither given over to gravity like the supporting limbs, nor yet quite freed from it like the head, but uniting the upper and the lower in the living rhythms of breath and blood circulation.

If we let our eyes travel slowly and reflectively over the human form or the human skeleton, from below upwards, they pass from a region of supporting forms with a downward gravitational tendency to a middle part, the structure of which mediates between the earth and the circle of the earth's environment, all of which is more particularly expressed by the arms and hands, until finally they reach an upper region which takes its stamp from the surrounding cosmos, from all that lies beyond the earth. We can watch the bearing power of the legs being taken over by the pelvis, and carried up through the spinal column with its elastic curves, see it flowing through the cervical vertebræ, until it is finally arrested at the place whereon the skull reposes. Arms and hands too are no less marvellously related to this upward flow; when they hang down they show, in their readiness to support or carry, their relationship to the legs; when raised horizontally, their relation to the domain of the chest and head begins to reveal itself; they put themselves at the service of the "higher" powers, and become organs of expression for man's inner being which lives in willing, feeling and thinking.

The head itself, by the form it has taken, has freed itself entirely from the forces of gravity; in its lofty repose it enters into connection with the surrounding world through the senses alone. Man's head makes it possible for him to stand face to face with objects because it does not itself deal directly with things like the limbs, but is raised above them as a kind of watch tower. All that man makes his own by the seeing, hearing and thinking of his head, he passes on to the middle system, informing with it the human speech that is saturated with his feeling and ensouled by his breathing, or he turns it through his hands into action, and with his feet makes it part of his "way" of life. Above, vision penetrated by the whole world; below, activity seeking the earth;

and in the centre the human pulse beating betwixt earth and heaven. Such is the description offered us by Rudolf Steiner's light-filled vision.

This picture, when considered by an unprejudiced, wide-awake intelligence is nothing strange. It points to the secret of man himself, to the essential difference between man and animal, and, going even deeper than Goethe's image of pillar and dome, gives the key not only to man's organs of contemplative reflection, but also to those of feeling and to those of work and action.

From this point of view it can also be seen that *all arrested development in man's body depends upon the fact that his upper pole is, so to speak, withdrawn from the gravity-forces of the earth.* The shortcomings of man's physical organisation correspond to his erect posture and walk. Because the head is freed from gravity, it is withdrawn from the tyranny of the moment's need and is thus elevated above the function of a tool. As the arms and hands were released from the necessity of carrying the body, they no longer needed to fashion themselves for the special tasks of the earth domain; and they were thus able to preserve their universality and to become at the same time limbs and organs of expression for the soul. The chest was able to attain to its human function as the bearer and transmitter of that speech, through which human feeling streams, colouring all its content for the very reason that human sounds are *not* uttered, like animal ones, under the urge of organic necessity. In man, a merely animal sound-production is held in check: his organs of speech have not evolved into a roaring instrument, as in the anthropoids, but have become the means of communicating nobler and more objective truths. Thus the whole of the upper portion of man's body bears the stamp of forces arrested, which hold back the animal development (see also p. 18).

The higher mammals, including the apes, are subject with their whole body to the forces of gravity. They have not resisted the fall into the horizontal. And thus it is that for them there is no "higher" system in the same sense as for man (Note 7). Their arms and hands (to name them according to their original character) are removed from their relationship with the higher system of the body. They have become instruments having to support the body and to adapt themselves for this purpose in every conceivable direction; they acquire hoofs, claws, talons; they become tools for running, swimming and climbing, and they take over the task of dragging, pushing, driving or hurling the body forwards (Note 8). From every such metamorphosis the human hand has been kept back; it has not become involved in the gravity forces, and for that reason has kept its original form. In this way, however, those forces of metamorphosis which expend themselves on the limb and chest organisation of the animal, bringing them to a state of physical perfection, are held in reserve in man. They can therefore enter his service in a far more direct way; and here already the reader gets a glimpse of something the full significance of which will emerge in the chapter on the "Soul." The formative forces thus kept back and in which the most varied forms lie latent, come to light again in man's creative imagination, and it is out of this that he fashions the tools which are lacking in his bodily organisation. Man's soul is equipped with all the creative power which has not flowed into the middle system of his body: and it is thus that he can provide himself with what he lacks—oar, shoe, climbing iron, aeroplane wing, and the whole range of his technical gear (cf. p.2).

In a similar way formative forces are withdrawn from the head of man. It has been shown in the foregoing pages how the facial portion in particular of the human skull as compared with the animal's, betrays a backward stage of development. Both the upper and lower jaws are arrested at an early stage; the teeth remain in fairly even rows, the eye-teeth do not grow into projecting sharp-pointed cones; no accretions or protuberances develop on the cranium itself; the foramen magnum retains its original position and does not move backwards; the head keeps its foetal shape of a regular even dome. In short, the whole "primitive" morphology of the head must mean a real economy of the formative forces and one which operates to the advantage of man's inner being. The head becomes the vehicle of intelligence and we may watch the progressive manifestation of this intelligence throughout childhood, while face, jaw and dentition are being " formed," that is, are being *arrested* at definite points in their development. With this Rudolf Steiner's statement becomes clear to us—that the formative forces withdrawn from the head are to be found again in the intelligence of the mature human being. *The imperfection of the human head* (as compared with the completed formation of an animal head) *has its correlative in the spiritual capacities of man.* What is physically thwarted comes to manifestation as the power of knowledge.

It is quite the reverse with the higher animals. The whole of the upper organisation, and the head with it, is in the grip of the forces of gravity. Upper and lower jaws hang like a great load on the head, which can no longer be carried upright. It hangs from the spinal column, or rather, from a spinal chain, and in this manner it enters the same category as the limbs and even becomes as described above, an auxiliary limb. We need only observe how the horse, for example, nods its head as it walks, and how with each one of its different paces, the rhythm and motion of the legs are imparted in corresponding motions to the head; and we shall understand that here the head has become merely the " front " end and is no longer raised aloft as the " head and front " of the body. We shall also understand what it signifies for the animal head to be bent down towards the earth and entirely subject to physical necessity; how the two eyes are quite unable to focus together on one objective, constituting it, in the act of doing so, an " object." Only when the head is relieved from implemental activities, can it raise itself freely to consider the surrounding world, and to mediate sense perceptions. Here too, it becomes clear to us, how significant is the retention by the eyes of their fœtal position at the front of the skull; without this no convergence of the axis of sight would be possible, and therefore, no focusing of objects. All these things man owes to the arrested development of his head (Note 9).

If attention is turned once again to the chest organisation together with the arms, it is noticeable that these really terminate above in the larynx, the tongue, the jaws, and the parts of the inner ear connected with them, which, as is well known, are formed from the branchial clefts of the embryo. Upon consideration of the interdependence of these organs we become aware of the physical part of what Rudolf Steiner called the speech organisation. This organisation is only possible in a body whose chest and arms are completely freed from the necessity of serving as supports, and whose trunk stands erect. Figure 2 shows this connection quite clearly; how finely and lightly the human chest is articulated with its contrast of upper and lower! The arms

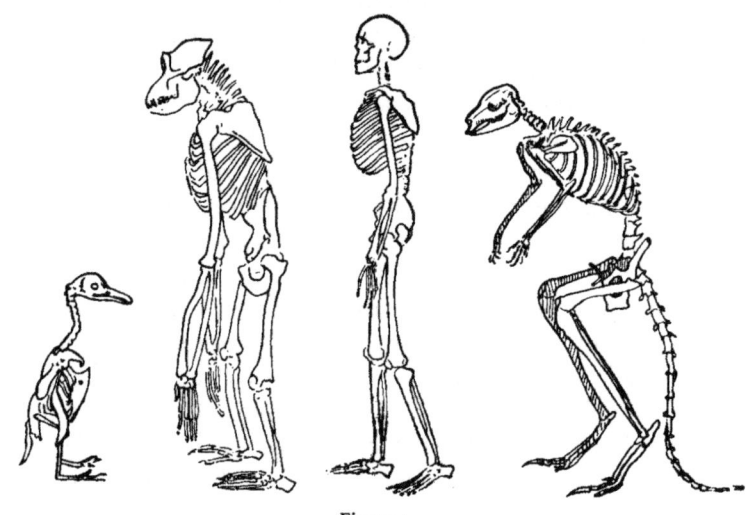

Figure 2.
(a) Penguin.   (b) Gorilla.   (c) Man   (d) Kangaroo.
(Only the human skeleton has taken the erect position.)
(a) After Eimer. (b) and (c) From Frizzi, *Anthropology*. (d) From Brehm, *Animal Life*.

appear to be only loosely attached to the shoulder-blades and are brought without effort into all those expressive attitudes and gestures whose language—immediately intelligible—supplements the audible speech which has first to be learnt. And now let us see what has become of this beautiful speech equipment among the animals; the arms are required either for support, or for running or climbing activities, or else they are deformed to hoofs, or lengthened or shortened. Among running animals the chest is contracted, while among the climbing ones it is covered with powerful muscles. The tongue is enlarged and hardened; the ear, distorted from its harmonious form, has become pointed and moveable, the lips, neither delicate nor mobile, are deformed to a jowl—only fit for snatching food; and here we clearly see why it is that animals cannot have speech in the human sense of the word, not merely for the external reason that their organs are incapable of it, but because these latter did not cease developing early enough to be ready to respond to other uses than those of physical necessity. The very thing which urged the chest and speech organs towards too strong a physical development in the direction of the tool, took from them the capacity to become an organ of the soul. Even where the external outline of the chest is curved almost humanly, as among the apes, the inner breast structure is far too massive, the arms are too heavy, the larynx has evolved much more as an instrument for making animal sounds, leaving no room for the development of speech organs. In particular, the larynx of the man-like ape shows a marked difference from the human; on each side there is a strong walled sac turned back all round from the inner side outwards; these sacs reach downward as far as the collar bone; and laterally almost to the arm-pits; are filled with air and serve as roaring-cavities for amplifying resonance. In place of these in the human larynx there are the above-mentioned (p. 14) diminutive crypts of Morgagni; these are not in any sense vestigial remains, but the germs of organs not yet developed. Once again in a few rare cases, these may enlarge, pointing us to an ulterior evolution which the anthropoid apes have actually carried out; and in such cases surgical intervention may be necessary.

Thus we have here a structure which, only in so far as its development is restricted, can serve man's speech-organisation. In the animal it develops into an instrument for magnifying sound. The growth which in man is only incipient, develops in the animal into a physical organ. In man the creative force thus economised is embodied in the treasures of his faculties of soul and spirit. The very thing that impoverishes him physically, makes him creative as a human being.

If the arrested growth of the upper system of man's body may be thus related in a rational way to the whole of his being, we are still faced with the great riddle presented by the difference between man and animal in the formation of the lower system. The reader will have noticed that the evidences of retardation just enumerated all relate to the upper and middle bodily systems. Does the lower organism reveal a different connection? Does yet another neglected law lie hidden here?

THE FOOT

Here again only unprejudiced consideration is of any use. Let us start with the human foot. Over and over again it has been compared with the hand; sometimes what was common to both, sometimes the characteristic difference between the two organs being subjected to detailed examination. The human foot is characterised by the way in which the vertical pressure from the leg is received by its arch, rayed out through it, and distributed to the ground. Most of this work is accomplished by the great toe and the metatarsal bone belonging to it. The arched foot is a characteristic human form, to be found in no animal; it is made up of the backward-facing heel, and the forward-facing tarsal and metatarsal bones; the toes do not belong to the actual arch but are merely attached to its fringe. The keystone of the arch of the foot is provided by the ankle bone, or astragalus, on which the leg and the entire weight of the body rests. If this structure from the knee downwards is considered, its wonderful morphological adaptation both for support and leverage becomes manifest. The human foot is a perfect image of its static task of supporting. It is evident that the legs are constructed as carriers, the human knee and thigh bone both keep the perpendicular position. The trunk rests on these parts of the skeleton, as if on two slender artistically proportioned columns; the knees are fully extended, bent neither forwards nor inwards. No animal on the face of the earth has this aristocratic distinction of the straight knee. Even in walking the nobility of the structure of the human body is revealed. The legs are true, even while in motion, to their pillar-like task of supporting, for the joints of this marvellous edifice bend only just far enough to allow them to become supports again the moment locomotion ceases. The bones of the arch of the foot are held together by very firm ligaments; the adaptable connection between shin-bone and foot, thanks to the service of calf-muscles, is a distinctly human formation. (Mephistopheles has to put on false calves when he wants to appear as a man.) Thus man's supporting limb-system is an organisation of wonderful perfection (Note 10). Nevertheless it is perfect in quite another sense than are those limbs which in the animal take the place of the human hand. And here we approach the consideration of a second great mystery in the structure of the human body. It is not easy to express in words because it is a great deal further from present-day conceptions of perfect and imperfect than that paradoxical perfection of the human hand of which we spoke before.

Let us start with the study of the hind limbs of the apes. It is well known that these have, in place of a foot, a hand-like structure. The toes have the shape of coarse fingers and are curved; the great toe is contrasted with the others as the thumb with the fingers. Only the heel actually suggests a foot and even this is considerably bent inwards, instead of projecting spur-like behind and below as in the human foot. It must be acknowledged that in the ape this organ has too much in common with the hand to be a perfect foot. If next we follow the ape's lower limbs upwards we get a still stronger impression of imperfection; the legs are much too short and too weak, there is no calf, and the knees are bent forward and inward. An ape running on its hind legs is really a pitiful sight (the man-like apes only do it for a few steps and the gibbon has to balance himself by raising his hands above his head). As the head and upper body of the ape give the impression of top-heaviness, so the lower limbs make the opposite impression. They are too light; they have not grown up to this task of bearing; they need the assistance of the hands—and in saying so, we have given the key to their deformity! The legs can only be developed as carriers when the full weight of the body is taken by them. Thus we are justified in saying of the ape's body that above and in front it is too heavy, and below and behind it is too light. The head merely " depends "; no image here of the surrounding universe: trunk and legs are not " the column supporting a vault destined to reflect the heavens."

Among true quadrupeds, the animals proper (ΘΕΡΙΑ) of the ancients, all four limbs contribute to carry the weight of the body and on that account have grown to resemble one another very closely—nowhere else but in man is there to be found a perfect example of the limbs achieving a position perpendicular to the earth. Even where there is a certain tendency to the pillar-form as in the free parts of a horse's or elephant's leg the trunk conceals a curvature of the carrying bones which points significantly to an incomplete mastery of the forces of gravity. Everyone knows that the backward projecting point on the hind legs of the horse corresponds to the heel in man, and that the rounded form to the front, higher up, at the actual spot where the leg leaves the body, represents the knee. This is generally forgotten when we see a horse in motion; otherwise we should have a weird impression of its running in a sort of squatting position, with drawn-up knees on unnaturally long feet, and always touching the ground with the middle toes only. (The bend of the knee is only just withdrawn into the trunk; and in camel and llama may still actually be seen.)

The form of a kangaroo or even a bird (see Figure 2) is almost more grotesque still. In all these cases the leg proper is too short and the foot much too long. In the bird the leg corresponds to the mid-foot—i.e., to one part only of man's arched foot which carries his weight. Even the stork does not stand on its legs, but on the middle part of its foot lengthened into a stick. A moment's consideration of these differences will suffice to convince us that only in man are leg and foot truly adapted to the static and dynamic conditions of the earth, and that in man alone the contrast between upper and lower, which is confused and hidden among the animals, has come completely to expression.

We could go on to trace this penetration of the nether organisation by the earth-forces higher up into the abdominal muscles, into the form of the pelvis, into the arrangement of the intestines, and so on into the curves of the spine (see Note 11). If we did so the law just set forth would be once more vindicated: that from above downwards the lower part of man's body is perfectly adapted to the sphere of the earth. The significance of this has been made clear by Rudolf Steiner as the result of his spiritual investigation, " the human limb-system remains during the whole of man's earth life—a germ which is prevented from developing." What form the human limbs-organisation *might* have taken is hidden from us, because of its compulsory adaptation to the field of gravity. The nether limbs of the human being are not the physical image of the formative forces in which they have originated. Instead, these formative forces have retired to allow full play to the gravity forces. Thus the limbs on which man supports himself owe their actual form to a complete sacrifice of the possibilities inherent in them.

While the upper pole of man's organisation bears the stamp of *retardation*, the lower pole is of the nature of a complete *disguise*. (The reality at its root is only revealed when man's earth life is finished; for then the eyes that can follow the human entelechy beyond earthly death see the hidden spiritual germ develop.)

The development of the embryo confirms the differences here described between the upper and lower organisation. The flattened extremities become visible as early as the fourth or fifth week, and display extraordinarily early a fundamental diversity. But while the upper limbs retain their flat shape and only divide at the fingers, those of the lower limbs form very early a heel and a true sole upon which the toes already show the stumpy form that characterises the human foot, as against the hind hand of the ape. Thus the foot of the human embryo is shaped early in a direction destined for *it* alone, while among the animals the difference between the two pairs of extremities remains inessential throughout their lives. The jumping and flying mammals and birds are the only creatures which present apparent exception to this. Here a significant difference appears between upper and lower, and either the fore limbs retain a certain similarity to the hand (in rodents, like the squirrel, jerboa; also in the kangaroo family) or they are specialised as instruments of flight (bats and birds). Thus a certain approach towards human semblance is significantly bound up with every elevation of the upper body from the ground; but with this difference, that nowhere, except in man, does the head really free itself from the force of gravity, and nowhere else are the nether limbs completely given over to the earth forces.

We have assembled these data in order to settle the physical antithesis between man and the higher animals by means of a comparison. The final result may be briefly expressed as follows:

THE TWO POLES

Man wrests his upper body free from the force of gravity. As a result the development of the head is arrested at a stage not much beyond the embryo. Arms, hands and chest are also affected by this retardation, and the form, stopped at an early phase, appears ennobled.

The animal surrenders the upper part of its body to gravity. It is true that its head begins development after the fashion of a human head, but it hastens towards a final stage in which it takes up into itself too many of the gravity forces. Arms, hands and chest draw nearer the earth and recede further from their original form the more they have to serve as supports to the body. The form thus completely metamorphosed appears "animalised."

Man submits his lower organisation completely to the forces of gravity with the result that it assumes the shape of a supporting pillar, appearing as a perfect organ in the realm to which it has been delivered over.

In the animal it is the opposite pole to the head which refuses to submit completely to the forces of gravity. This brings about an excessive similarity between the two poles of the body. The animal is too heavy where it should be freed from the ground, and too light where it should develop weight-carrying capacity. Thus its upper organism cannot keep close to its archetypal form, while in its lower structure it is but an imperfect expression of the earth, whose shaping forces it will not fully admit into itself.

Man, in the upper pole of his organism, retains a past condition, which the animal has abandoned.

He strives, in his lower pole, towards a connection with the earth, to which the animal is denied access.

PART II

DESCENT

Man is the lord of creation by right of primogeniture.
—K. Snell, 1863.

INDURATION

Living beings achieve maturity through a process of hardening; the plasticity of the youthful body gives way before an ever-increasing firmness. The final form emerges from the prototype: it has lost all shaping power: it is *finished*. This rigidification is a necessary condition for higher grades of work and for the most vital activity; but at the same time it obliges evolution in its further progress to return at the end of the individual existence to the plastic condition present at the beginning. In the plant this is exemplified in the bud on the stem, or the germ within the fallen seed; among the lower animals in the new individual dividing from the body or hatching from the egg; among the higher animals and men in the embryo growing in the mother's womb. The final form, as such, is always infertile; it must either wither or become a mere sheath to be discarded by the newly emerging life. *For progress and propagation always mean a recapitulation of the stage of immaturity and imperfection.*

This law underlies the formation of every organ. The preceding chapter showed how it applies in the case of the extremities. Nor is it difficult to demonstrate that a finished tool must be an end-point, beyond which no further advance can be made. New organs can grow only from budding and indeterminate conditions. Their development involves a continual *abandonment of potentialities*, a repeated disjunction of cognate growths, and a final concentration in *one* direction only.

Doubtless it was the same law that determined that gradual

evolution of living Nature which the past conceals. Evolution resembles the plant which, springing from the seed of primeval epochs, attains a more and more majestic development as its branches multiply. The oldest forms of life, germlike and inchoate, must yet have borne within themselves the whole infinite range of possible developments; but among their descendants a division will have taken place, some departing this way, others that, at the expense in each case of their universality; every advance step must have been purchased by an irrevocable fixation.

A trichotomy of this nature governs every visible process of development: the originating movement towards differentiation and determination brings in a state of perfection, of which in its turn the inevitable consequence is a certain rigidity.

In every case once the separation had begun and something new was trying to free itself, a form which had remained undeveloped, a bud-like form, had to supply the starting-point. *New forms in the earth's history could arise only from the " eyes " or dormant buds on the great trunk of creation.*

If, therefore, we search seriously in the history of the earth for the progenitors of a group of animals, it is to the undetermined and plastic forms that we must look. And if we desire to picture these, in no case must we turn to the form and structure of the " finished " animals of to-day; for these latter, primitive and highly organised alike, have yielded to the desiccating process. The new creature, pregnant with the future of the race, could never spring from such an end-point, but only from some seminal condition more manifold and versatile in its nature.

It was by a singular fate that the evolution theories of the nineteenth century left all such considerations entirely on one side. Yet, a kindly fate withal! We have only to think what might have become of the theory of descent, had a sceptical attitude of this kind ruled from the outset. The extraordinary achievement of outlining the course of organic evolution would have been impossible! The structural plan of the theory of descent had to be laid before such questions as the above could even be put. The pioneers of that theory, therefore, have earned the lasting credit of laying the foundations of a great advance in knowledge. With the utmost courage, with the widest possible survey of what was then a welter of chaotic facts, they affirmed the idea of the gradual evolution of man from primitive animals, and lost no time in plotting that idea on their charts of descent—a marvellous achievement for all time! The inimitable penetration with which Ernst Haeckel in particular indicated the minutest details in his genealogical trees and thus led on the theory of evolution to victory cannot but win him the respect of everyone who concerns himself with these problems. It is easy enough now to blame the overhastiness of those days; but without that devoted and enthusiastic band, the general outline would still be lacking.

MAKESHIFTS

Thus to obtain anything approaching an all-round view of the path of evolution, we are justified even to-day in doing what Haeckel did, in relying on analogies with the forms of living animals. For if we would reconstruct mentally the series of progressive forms, we are obliged, in order to save our ideas from fading away altogether, to turn to the already known. This is not the only sphere in which thought has to be grateful for examples which afford passing help only. A man wishes to get a concrete idea of the plant: he thinks of a rose; to represent the whole natural order of the worm to himself, he thinks of an earth-worm. Thus each investigator seeks to get a clearer idea of the ancestral chain of man and animal by relating it to the present-day representatives of its various stages. Of course when he has actually constructed his picture of evolution, he should remember that all this is only a help to his imagination, and *that a genealogical tree peopled with contemporary forms cannot reproduce the actual facts of evolution.* If he forgets that its original object was illustration, it will simply blur his vision.

That the animal forms living to-day could not be the true ancestors of the species animal and man, is easy to understand, and has never been disputed. But the question becomes infinitely more complicated when we approach the fossils. Palæontological research has expanded marvellously since its findings were used for the genealogical tables. And here too, it must be admitted that this advance would have been unthinkable had men allowed themselves to be seriously disturbed by doubts. In fact, it was accepted, tacitly but none the less as a matter of course, that the " ancestors " of man *must* be somewhere to be found among these deposits; a piece of ingenuousness which in its time was all to the good. To-day, however, when palæontological discoveries are invariably viewed in the light of the evolution idea, it is time to ask ourselves carefully of each form which has been appropriated to the ancestral class, whether it really fills the part assigned to it. After detailed study, we can hardly help feeling about these extinct forms the same uneasiness which the contemporary ones arouse. The whole physical structure of their remains, the very way in which they have been preserved, shows that they derive from beings which could not have been (in the literal sense) *materially* different from those of the present day; they are worked out with the same attention to detail, highly developed and " finished," even when in certain anatomical particulars they reveal so-called primitive characteristics. Their armour and their skeletons are elaborated with that wealth of detail which we expect in fully developed creatures. Their finely chiselled perfection (consider the armour of a Cambrian trilobite, a representative of the oldest fossil-bearing stratum!) has nothing of the embryo about it. These creatures were no less mature in structure than any animal now living. Can the parent stock really be hidden among such forms as these? Or does that not involve imputing a flexibility to them which their bodily forms prove they never possessed?

Fundamental considerations of this kind are, as we said before, overdue. And it is the apex, surely, of the genealogical tree which will have to submit to the most substantial correction of all, for man's position at that favoured spot is, as the first chapter showed, being seriously shaken. When once it is recognised that the human body has retained to this day indications of a previous and inchoate stage, then we begin to understand how just that body itself points us far more clearly to the true ancestors of man and of the higher animals than any creature out of its environment. In the nineteenth century men placed the human form at the summit of the animal kingdom, never dreaming of the difficulties that such a view would create. Yet this ingenuousness had its advantages. A table of the origin of man and animal which ignores the essential point of the " primitiveness " of man will no longer hold water. The original form of the higher animals must undoubtedly have resembled the human form more closely than it resembled our animals of to-day. In this connection too, therefore, the genealogical trees need a severe overhauling!

The alterations to which the prevalent ideas on the subject must submit are clearly drastic. Scientific research is faced to-day with the alternative of definitely abandoning the theory of evolution or helping it to take a radically new form. The age stands before a decisive rebirth of the whole theory of evolution. We should make up our minds to assist at this metamorphosis, unless we wish to sacrifice the progress which the natural scientists of the nineteenth century achieved, and to render their immeasurable contribution vain.

★   ★   ★

THE RECONSTITUTED TABLES OF DESCENT

Let us begin by indicating the direction in which this metamorphosis must happen. A careful study of the facts makes it clear.

Take the genealogical tree of the vertebrates reproduced here from a sketch by Haeckel. The essential point in older representations of this kind can be seen at once: it is that names of animal groups now living (such as selachians, amphibians, pouched animals, lemurs, anthropoid apes) are confidently placed at the forks on the main stem. Clearly such a diagram shows no actual genealogical tree; it merely brings together the results of the researches of Comparative Anatomy. That this can be done in the form of a tree at all is itself a proof of the fundamental rightness of the theory of evolution. Undoubtedly, there is a foundation of reality in this method of putting types together; but it must be remembered that it originates simply and solely as a *picture* indicating which types the ancestors of the vertebrates probably belonged to—or, through what stages of evolution they have passed. In place of the type there is inserted the name of the animal group which represents it to-day.

The only serious difficulty is this: we have to take it for granted that a creature, let us say of the type of selachians—that is of the present-day sharks and ray—had among its descendants certain species which have evolved through all kinds of varying circumstances into the higher fishes (teleostians, etc.) of the present time;

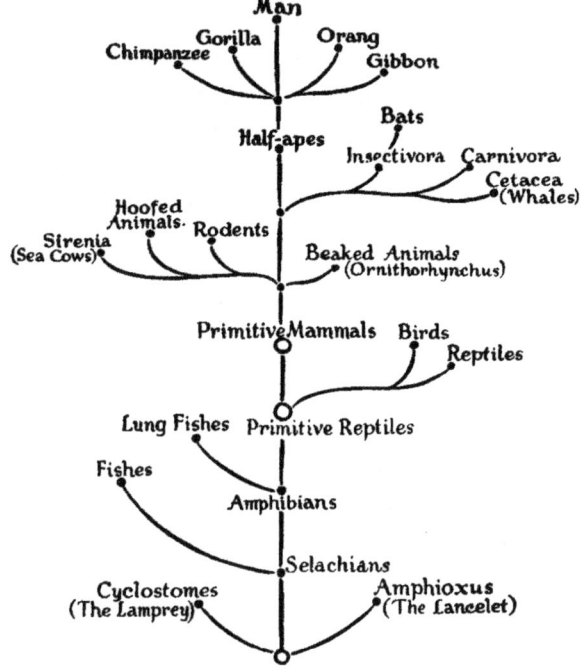

Diagram 3.—Haeckel's genealogical tree of the vertebrates.

others which have developed into the amphibian species and therefore have become the ancestors of the present land animals; while a third group from that time to this has remained at a standstill, at the stage of the shark and ray of the present day. Similar assumptions must be made for the type forms occurring at the " forks " higher up in the diagram. We have thus to presuppose as ancestors creatures which, in spite of pronounced and definitive organic characteristics, possess a plasticity which enables them, in the course of vast periods of time, to produce new types from themselves.

Yet among the descendants of these surprisingly adaptable progenitors there must needs have been others characterised by an equally unheard of rigidity, a rigidity which has made it possible for them to stick at their original stage down to this very day. We begin to see that either this adaptability must have depended on quite different earth conditions—in which case we do not understand why their contemporaries stood still; or the elementary conditions of ancient epochs resembled those of the present day—in which case the adaptability of the first group is equally unintelligible. Two such absolutely opposite qualities occurring in descendants of the same parent stock are inconceivable; and it simply will not do to shift the burden of explanation on to the environment, for in one case this would have to have worked continuously and progressively for thousands of years, while in the other it effected scarcely any change whatever.

THE ANCESTRAL FORM REMAINS AN ENIGMA

The first glance at any such table teaches us, then, that the ancestral forms remain completely enigmatic to our present-day conceptions of Nature's creatures. We form a better idea of them if they are imagined, not as concrete beings filled with material substance within a world such as ours now is, but rather with the bareness of those textbook diagrams, which, lacking any distinctive character, are susceptible of particularisation as required. If we ascribe to an ancestral form an organisation adapted to particular earth conditions, we at once find it impossible to reconcile this with the fact that it is supposed to have included, nevertheless, the possibility of putting forth new side-branches. If, on the other hand, we grant the qualities of a genuine primitive form, namely, indetermination and adaptability, then once more it is incomprehensible how such a primitive form can have " lived " in a world like the present one. Turn it how one will, the nature of the ancestral types remains the real problem in all theories of descent.

The reader who trusts to the theories still valid to-day may here object that palæontology has already reconstructed a great number of "intermediary stages" from the organic remains found in the earth's crust.

It is worth while considering whether these discoveries can really fill the position assigned to them on the genealogical tree (see Diagram 4).

Neanderthal Man (*Homo Neandertalensis*), discovered in 1856, represents a stage of archetypal humanity, which has since been further exemplified by dozens of skeletons. Since Peking Man (*Sinanthropus*) is clearly an oriental parallel form of the same age, we must assign a world-wide dispersion to this "primitive"

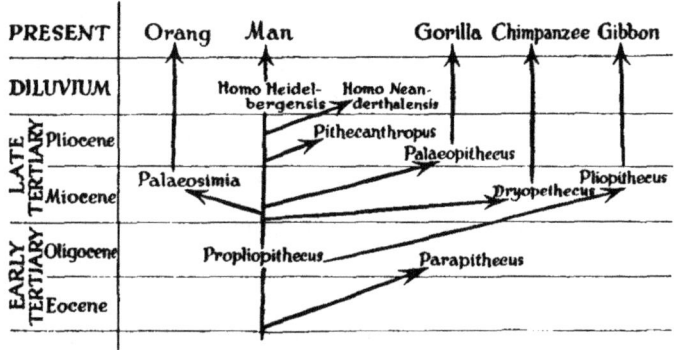

Diagram 4.—Tables of descent, showing final points according to Gregory.

stage. The basic tokens of humanity are unmistakeably present: the upright gait, the lower jaw with its half-circle of regularly formed teeth retreating contrary to the cranium, the human formation of the hand and arm, and the sustaining limbs adapted to gravity (see Chapter I). These basic tokens are instantly recognisable, but along with them go certain characteristics of early man which clearly have a very different significance and which determine his "primitive" appearance: the low forehead, the protruding eyebrows, the small brain capacity, the thickness of the skull, the coarse lower jaw, the small size of the whole head. At the same time it cannot be doubted for a moment that it is a human form we are concerned with; for these people are tool-users and acquainted with fire. Along with Peking Man were found thick ash-heaps, which prove that he kept his fire in for days—or probably weeks—at a time.

Nevertheless, Neanderthal and Peking Man are not regarded by most authorities as the direct ancestors of modern man, but as "blind" side-shoots on the tree of human origin. The thickness of the skull bones, which are found, by contrast with to-day, even in youthful skeletons (the walls sometimes a centimetre thick!) can be taken, literally, as examples of "persistent induration."

Pithecanthropus, discovered by Dubois in 1891, is harder to place. He figured at first as a half-human form intermediate between ape and man. Now that several much more complete fragments have been found in Java, hardly any doubt remains that what we have here is a human being of very great antiquity. (On grounds of priority we have to retain the *name*, though it is no longer appropriate.) "There was only one very small piece," writes von Königswald (1955, p. 119) "missing from the roof of the skull . . . and at the back the cranium was intact up to the edge of the occipital hole. Above all, the temporal area was well preserved. Now it is just the part round the ears which is of decisive importance for the question—man or ape? With the ape the ear lies more or less on the prolongation of the zygomatic arch, but in the case of man lower down. With the ape the cavity of the maxillary joint is shallow—in a man it is deep set. Our new discovery shows clearly that it could only be a human skull; thus it was proved once for all through this discovery that the controversial Pithecanthropus must have been a man."

Now more recent geological investigations on the site have shown that the stratum of Trinil to which Pithecanthropus belongs is not in the Tertiary but in the Diluvial (Pleistocene) layer. This proves that Pithecanthropus was not a predecessor of diluvial man, but a contemporary. By the same token he cannot be an ancestor of Neanderthal Man. Thus, we see that on the most modern "trees," one of which we reproduce in Diagram 4, Pithecanthropus, as well as Neanderthal Man, is relegated to a side-branch. It is significant that this corresponds with the almost universal opinion of modern palæontologists. Once again, on the genealogical tree itself there is a conspicuous vacancy (Note 12).

What of the additional and possibly still older forms which have been added in the last few years? Some of these such as Meganthropus (Gigantic man) described by von Königswald, undoubtedly belong to a primitive human type. The fragment of lower jaw—in spite of its " improbable size for a human jaw " —is nevertheless in fact human. The back premolar has only one root, though in all the apes it has two; while on the inner side behind the chin the insertion of the so-called *spina mentalis* is visible. This insertion is particularly significant because there are attached to it certain lingual muscles which serve the human being in the production of articulate speech. (Peking Man also has this " spina.") Meganthropus, then, definitely possessed the faculty of speech and so was a genuine human form.

We may pass over here the teeth, described by von Weidenreich (1946), of other " gigantic " types from China, since nothing further has been ascertained about the possessors of these strange but definitely human teeth of enormous size. All the same it is interesting to recall that Rudolf Steiner spoke of gigantic human forms, which afterwards died out, in connection with the " Atlantean " (Tertiary) epoch (see *Geheimwissenschaft im Umriss*, Dornach, Ch.4. English translation: *An Outline of Occult Science*, New York 1923, and later editions).

All these, then, are early human types and present a certain contrast to the species exhibited by Broom, known as Paranthropus and Plesianthropus, from Africa. Paradoxically, the place of the latter in the ancestry of man is rendered uncertain by the fact that their teeth are graded in a most unusual way. The incisors are too small, the molars, on the other hand, too large for humanity. Clearly these species have taken an evolutionary side road.

As to the *Australopithecines* from South Africa, it is true that they are located near Meganthropus (see above), but they fall outside the human kingdom. The reason why young specimens, such as the well-known Taungs find, look more man-like than older ones, has already been explained (p. 8). Their development leads them to the brink of humanity, but then falls away again (compare von Königswald, 1955, p. 167). On the other hand the latest types to be added to the list, Proconsul, Dryopithecus and Sivapithecus, indicate a definite direction towards the ape group

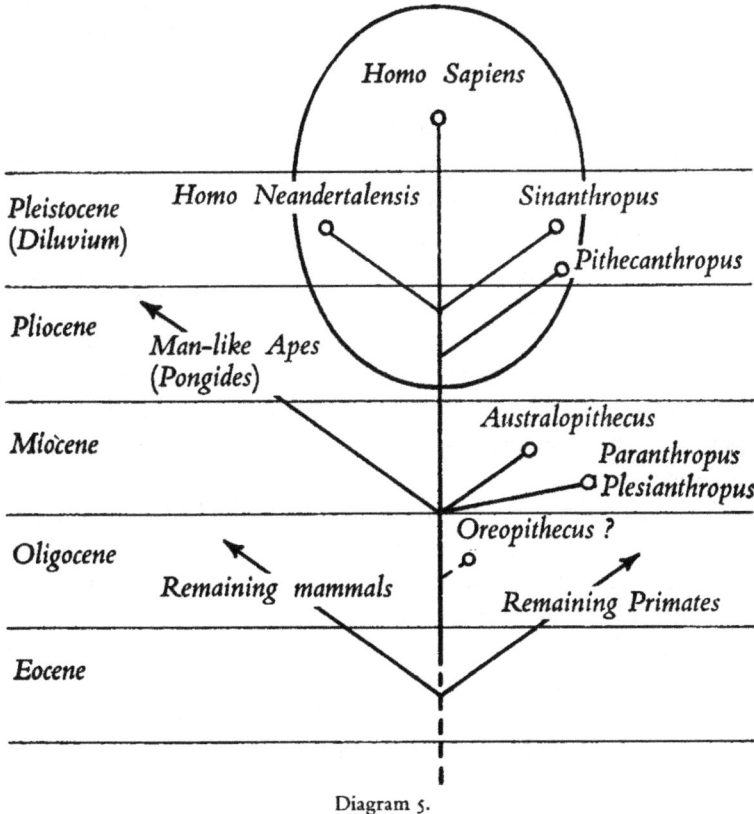

Diagram 5.
Branch-tips of the tree of human descent, revised in accordance with the state of research at the present day (1958).

proper. In no case are they ancestors of man (compare von Königswald, p. 149 ff.).

In every instance up to now a thorough examination of the fossils put forward as " stem " forms has yielded the same result. Only a very few of these fragments are worth considering as direct ancestors of man and those only as long as they remain imperfectly known. As soon as details are mastered, they belie the role originally assigned to them; every form turns out to be too specialised, too one-sided, to fill the part of a genuinely ancestral species. Thus, in the course of the years, the old, heavy-laden genealogical trees have been substantially *thinned*. More and more forms are relegated to the side branches, and the trunk grows bare.

THE INADEQUACY OF THE FOSSILS

Is this failure on the part of the fossils merely bad luck? Bad luck continually recurring? Can we go on for ever making " the present incomplete state of our knowledge " an excuse? Granted this paucity of material, still why should the gap occur every time just at the decisive place? Have we really any right to go on comforting ourselves with the hope that some later discovery may at last reward the expectations which have hitherto been so often and so bitterly disappointed? or would it not be better to abandon this hope once and for all and to draw a bold inference instead?

Such an inference can only be the one which has already been vindicated (see p. 33, above). The *real* ancestors of man and animal, with their protean morphology, are for us—with our present-day ideas about living creatures—structural enigmas. *They have left no traces in the record of the rocks.* It is the same problem under two aspects—the remains found in the earth's crust are not those of our ancestors: and the bodies of our ancestors have not been preserved in the earth's crust.

THE ARCHETYPAL FORM

It is of course comprehensible that an investigator of the present day should be unable to bring himself to accept such an unusual hypothesis without a struggle. It gives him no foothold within the range of his customary ideas. Yet he has only to review all previous attempts to complete the ancestral tree downwards, to be driven back into the very arms of the unwelcome notion. The necessity for supposing plasticity grows more and more imperative, the lower the position on the tree to which the form under consideration is to be allotted—to imagine an ancestral form possessing even the meagrest attributes grows more and more difficult the farther back one goes in time.

Even of the apes the ancestral type is uncertain, since these are supposed to include western apes and eastern apes, as well as the half-apes which differ from these so greatly. How nebulous this makes the original type can be seen from the great variety of opinions which obtain among palæontologists and morphologists. This uncertainty is greatly increased when it becomes a question of indicating the place at which the group of primates (man, ape, simian ape), or rather their forefathers, separated from the ancestral stem of the mammals as a whole. Gregory, for instance, derives the primates from arboreal insect-eaters, that is to say, from the hedgehog family. Huxley had already accepted the hedgehog as a primitive " collective type." But in order to trace the culmination of the human genealogical tree, back to a hedgehog-like origin, Gregory has recourse, not to the hedgehog itself, but to its tree-dwelling relatives, the tupaia (principal genus of Tree-shrews); and moreover, he has to consider these as relatively " large brained." If we look at all this without prejudice, we can see how a given form is tinkered with, how the most contradictory feaures are grafted on to it, and all to one end—to enable it to occupy a predetermined position on the genealogical tree.

The palæontologist falls into a quandary of this kind each time he reconstructs an ancestral type. If he tries to depict " the primitive mammal," his embarrassment becomes even more painful. Haeckel simply put that vague appellation itself at the appropriate place on his table of descent (Diagram 3, p. 32). But the more minutely we go into the structure of any such type, the more completely the attempt breaks down. At first of course it is natural that the original type should be assigned to that group of pouched and beaked animals (marsupials and monotremes) which comes low down in the system of mammals, for the former have " as yet " no placental formation, and the latter lay eggs as do the reptiles, which they resemble more closely than other mammals in their inner structure. But more exact investigation leads to the usual attitude of resignation: " Let us state once more with the utmost emphasis," writes Schwalbe (1916, p. 270), "that the pouched animals, not to mention the monotremata, have no connection with the evolutionary stages of the placentalia; they are *divergent* forms, and, as such, have no bearing on the problem of the latter's descent." It is always the same. *The form which would fit the pedigree is not known, and the form which is known, does not fit the pedigree.* It turns out to be "divergent," i.e. on a side branch of the tree instead of on the trunk. And the real ancestral form withdraws once again into its accustomed darkness.

The same truth could be demonstrated in the case of the reptiles and amphibians. The connecting link is too nebulous, and on a closer study the thin thread breaks. The broader the tree of descent becomes below, the easier it is to make sweeping assertions about the knot whence this or that side branch "manifestly" springs. Almost all the authorities concur in deriving the vertebrates at some point or other from the worms; but this huge order itself includes so many different types that no single characteristic can be found common to them all; it is simply a sort of lumber room. So that the statement as to the genealogy of "vermiform" creatures is no more than a meaningless truism.

Hence it was that as early as 1887 the wise Karl Snell could make this arresting observation: "It is quite useless to say that the ancestors of the vertebrates were worms, because such a statement has no meaning whatever. We should first have to define what we mean by worms; and no one can do that, because all that is known about them, is what they are *not*. The answer that the ancestors of the vertebrates were worms tells us nothing that was not known before the answer was given, viz., that these ancestors were neither radiates, nor arthropods, nor molluscs."

THE LACK OF TYPE-FORMS AT THE FORKS

It is because so many investigators have had a similar experience during the last few decades that the genealogical tree has altered, as it has, in a definite direction. Originally it was always the actual name—of a group, of a fossil or of a hypothetical ancestor—which stood at the great forks in the genealogical tree. But one after another as the bearers of them became better known, these names were thrust out on to side branches, *the name of the proper axillary type being replaced by a note of interrogation*; until at last every actual genus had been removed on to the side branches and the main trunk was nothing but a chain of queries.

Thus the branches of the genealogical tree grew more and more luxuriant, and the main trunk barer and barer. Today it is obvious that the tree is no longer alive, no longer able to nourish its own branches. At any rate no one can now say exactly *what* it is that is supposed to flow as living sap through all the unnamed branches and twigs to the densely populated tips. The whole subject of descent threatens to sink back into its former category of the " unknowns."[1]

Another significant and even more extensive change has taken place in the genealogical tree during the last decades. The more scientific research came to recognise the gulfs that yawned, in respect of their structure, between one great animal group and another and the extraordinary divergences among the smaller subdivisions, the more it tended to thrust back the side branches of the genealogical tree to the earliest possible epochs, i.e. to give to these latter starting-points as low down the tree as possible. The reader can convince himself of this characteristic alteration by comparing the genealogical tree of the crab order, as given by Haeckel (1873) with the same as given by Giesbrecht (1912) (Diagram 6). Even without a closer acquaintance with the individual groups themselves he will see at once that the names which Haeckel places either near the point of origin of the entire genus or else at one or other of the many " knots " in the tree are now, forty years later, placed without exception at the ends of the branches. Let him notice especially how a single group, such as the Leptostraca, is taken in the modern tree right down to the original (and hypothetical) crab form, while in Haeckel's diagram it is first found above the fork marked Phyllopoda; which same Phyllopoda are themselves in the new diagram put at the tip of a branch. The result is that the whole tree is turned into something more like a bush; and it is questionable whether the newly invented labels (Protostraca and Protomalakostraca) can do much

[1] Translator's footnote—Thomas Carlyle's word.

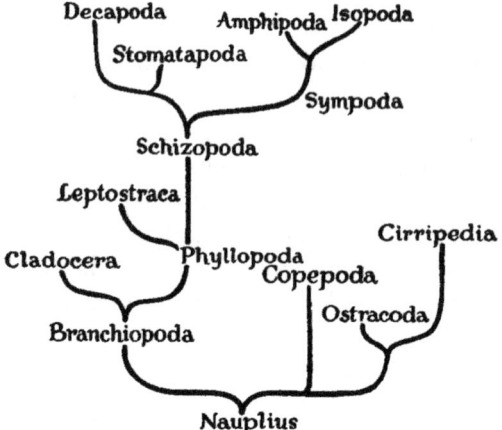

(1) From Haeckel: Natural History of Creation, 1873.

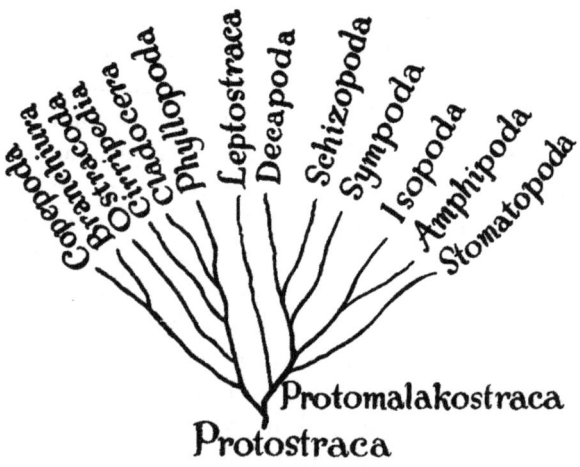

(2) From W. Giesbrecht: Crustacea, 1912.

Diagram 6.—Alterations in the table of descent of the crab order in the last forty years.

to preserve the system from its violent propensity to disintegration. A very little more and the individual branches will be parting from one another entirely and springing up side by side out of unknown soil! Now a system like that does not require any theory of descent to justify it. The idea of the branches springing up side by side and independently from an unknown source had already been used by Linnaeus in his Systema Naturæ.

However, it is not for the sake of cheap criticism so much as in the hope of fresh knowledge that we should concern ourselves with these difficulties. We should regard the lack of authentic remains in the geological records less as opportunity for destructive criticism of the theory of evolution as such, than as a spur to further and untiring reflection on the true nature of those hereditary forms through which the forefathers of man and animal must have passed.

MAN'S PLACE LOWER DOWN ON THE TREE OF DESCENT

Towards such reflection upon the condition of our lineal ancestors the studies undertaken in the first chapter afford unexpected assistance. In the present chapter we have not up to this point availed ourselves of this help, and many readers may well have been looking out for it with impatience. It is quite clear that the phenomenon of man's arrested physical development which was there demonstrated compels the conclusion, *that the human figure of the present day can give us a better idea of the original form of the higher animals than any existing animal.* If man has " retained " better than any animal his " primary " characteristics, such as the five-fingered hand, then his name ought really to stand at a much lower place on the genealogical tree than Haeckel proposed for it. If the habit of seeing this name at the top of the tree were not so deeply ingrained in us, it would long ago have been moved farther down, to somewhere about the place which Haeckel gave to the " primary " mammals, the pouched animals! Up to the present no scientist has ventured to sketch any such table of descent, though many already have at least advocated putting the human branch lower down the ancestral tree and making it diverge quite near the one appropriated to the other mammals! They have thus conceded to man the capacity for a development continued through long ages, but while his axil was removed to a lower place, his name remained right at the top above all the animals. In view of the importance of this subject, we may offer documentation of a kind which is calculated to bring this downward movement before our eyes.

Klaatsch on the 26th July, 1900, at a meeting of the Prussian Academy for Natural Sciences made the following statement: " The apes now living represent forms that have evolved in one-sided ways and are partly degenerate. The less an ape has deviated from the original type, the more man-like he appears. . . . Thus we can speak of a connection only at the root of the common genealogical tree, and this holds good finally for all mammals . . . *according to my idea man is a form central to all the mammals and primates* " (Klaatsch's italics).

M. Alsberg in a small book on the Descent of Man (1902) comes to the conclusion, "That for man there seems to be no question of an ape origin; we must rather *presuppose descent from a much more remote point* on the great tree of heredity."

The anthropologist C. H. Stratz writes in 1906, " The true beginning of man, i.e. his attaining to the upright walk and the beginning of a more vigorous brain development, I would place at the very least as early as the late Cretaceous period. In the oldest Tertiary era, which follows directly on the Cretaceous, we already find present the whole series of higher mammals. Since man is of a more ancient lineage, he too must have been in existence at that time."

Thus the Cretaceous period is put forward as early as 1906 as the epoch of man's appearance on the scene—a conception of evolution diametrically opposed to the usual one!

The well-known palæontologist E. Dacqué, whose book *Urwelt, Sage und Menscheit* (*The Prehistoric World, Legend and Man*) appeared of course much later (1924), goes even further when he writes: " Among the vertebrates the structural form of man takes us right back to the original amphibian stage. This fact gives us a new standpoint from which to estimate the antiquity of the human type or family. . . . We may expect to find the *human family*, as such, as early as the old-Mesozoic era, indeed even in the late Palæozoic era: to find, that is, an entelechy distinguished from the rest of the animal world by the possession of essentially human qualities, namely certain psychic and spiritual attributes " (see pp. 73–4).

Of course all these investigators know very well the responsibility they take upon themselves in inculcating this idea of the great antiquity of the human race: it opens the way for a point of view which strikes hard at the current theory of evolution. For the orthodox system of an upward evolution of species, under pressure of the struggle for existence, requires to complete it, the human form at its summit. The moment we admit a vast geological antiquity for the human form, it becomes impossible to ascribe the arrival of man to the pressure of the struggle for existence. What powers they were which sustained him in his archaic form must then remain a complete riddle. Klaatsch then is quite consistent in saying that man owes his form to the circumstance that he has been " spared the blessings of the struggle for existence. . . ."

It may be, however, that beneath the utterances of these investigators there lies a still more far-reaching divination, a notion which bursts the bonds of current theories in a yet more revolutionary way and thereby wins through to a still bolder survey of the facts. This idea is that the evolution of organic beings can in no sense whatever be thought of as existing apart from the evolution of man, that their progress is causally bound up with that of man! In which case it would be the mysterious thread of man's own ancestry which had provided the root stock of the great tree of creation, and which alone had transformed the latter into one homogeneous system, supporting it from top to bottom after the fashion of an axis. From the generations, as they struggled upwards, step by step, and under ever new conditions, the animal world would have separated after the manner of side branches diverging from the main trunk, branches whose existence would be unthinkable without the life surging through the trunk itself. The animal world would thus be the by-product, the waste-product of advancing human evolution.

THE NECESSITY FOR A NEW CONCEPTION OF EVOLUTION

This is that conception of man's "primogeniture," which the genius of K. Snell proclaimed as early as 1863 (Note 13). For in Snell's presentation evolution was still upheld by the spirit, and it was still possible for this idea to find a place in it, whereas from the mechanistic conception it has completely disappeared.

Here again precisely the same riddle as before confronts everyone who would approach this new knowledge, i.e. the question as to the material constitution of those predecessors, who were the first parents both of physical man and of the animals in the world to-day. For there can no longer be any doubt that these ancestral figures cannot have possessed the substantiality of present-day animals and man. They must have been creatures of an incomparably greater plasticity. If we admit this, however, we must reject once for all the notion that the evolution of species can have run its course in a world resembling, *in the solidity of its constitution*, that of the present earth existence.

On the other hand it is not sufficient to say in a general way, as is often done, that the physical and chemical conditions of the oldest earth epochs must have been "favourable." We must turn back to a series of telluric conditions which go far beyond the ordinary conceptions of Natural Science, if we are to form any adequate picture of the "becoming" of man and his companions on the earth. This picture, if it is to reproduce truly the events of the past, must be able to show how the separation of the various kingdoms and types took place; how, in particular, the phenomenon of retardation in the human form, came about: finally, how it happens that when man first appears geologically, he brings with him from the outset an organisation essentially human. A new picture of evolution, a new description of the past conditions of earth and humanity is necessary.

★ ★ ★

RUDOLF STEINER'S CONCEPTION OF SPIRITUAL EVOLUTION

Such a picture must needs be rooted in a deeper insight into all that is "growth and becoming," in a direct comprehension of the forces of growth, not in an external comparison of things that have reached completion. The cosmos, including earth and man with all the creatures, must be comprehended by means of a power of vision able to penetrate to the hidden phases of its development: then the resultant idea must be communicated to the ordinary mind, in intelligible language. It is self-evident that the older the conditions described, the further behind us we shall have to leave those conceptions to which we are accustomed—*and this precisely to the extent that the description is true.* The delineation of past eras, into which natural scientific investigations cannot penetrate directly, *must* of necessity sound strange in the first instance; it will demand a certain effort from the reader or listener. In return, however, a rich gain in knowledge beckons him on precisely at the point where contemporary science is inevitably held up.

The presentation which meets these demands is to be found in Rudolf Steiner's writings on Spiritual Science. It is based on his own insight into processes and things of the world inaccessible to our ordinary senses. He himself named this mode of knowledge Spiritual Investigation; he showed the way to it, and left us its results in the form of a comprehensive system (Anthroposophy) of a new knowledge of man and of the world.

The way in which this spiritual scientific system of evolution[1] answers the questions put forward at the beginning of this chapter, and many others also, is amazing. That the reader himself may acquire the necessary insight, the vast picture must be reproduced here in broad outline; it can then be confronted with all the unsolved problems in the theory of evolution. Everyone can satisfy himself of the fruitfulness of such a proceeding—though of course all our prejudices throng to oppose it—and each one of us can ask and answer new questions for himself.

[1] First published from 1904 onwards in the essays " *Aus der Akasha-Chronik* " in the monthly magazine *Lucifer Gnosis*; reprinted in the weekly paper *Anthroposophie*; Years IX and X, Stuttgart, 1927-28 (English translation: *Cosmic Memory: Pre-History of Earth and Man*, New York 1959). For a further description see *An Outline of Occult Science*, chapter on " The Evolution of the World and Man."

OUTLINE OF THE EARTH'S EPOCHS

The evolution of the earth, as it is seen by supersensible perception, begins amid quite other conditions than those which Natural Science can describe. *Everything that comes into being, arises from a spiritual source.* At the starting point of every event stand purely spiritual origins. It is from the deeds of Spiritual Beings that evolution proceeds.

No one should say, he does not understand this; it is the most self-evident truth in the world; perhaps the only one which can be understood by the human race as a whole without further explanation. For everyone understands what it means when he *does* something. He acts, and in the act perceives quite clearly the difference between the being who does something and the thing that is done. The decision to act is the affair of the agent—is that by which the being takes hold of its body and moves it on to fulfilment. No proof is needed to confirm this; it is self-evident. It is much more fundamental, more "given," than the world itself. The deed always comes first. It is the origin of all fact.

Thus the world of the spiritual, in which all events have their origin, is a world of active beings, filled with creative power, filled with the impulse to become, though in the first place still without manifestation. Just as the sense-world consists of things, so this original spirit-world consists only of beings; they are, so to speak, the material of which it consists. Innumerable, separated into different ranks according to their aims and the range of their activity, nevertheless they co-operate with and support one another in their work. The highest in rank give the impulse; the next shape it, the lowest minister to it.

Rank upon rank, reaching one beyond another and working into each other's hands, they cause the "Universe" to come into being. Time and space take form. Gradually "realising" itself in mighty rhythms, standing out from its source, consolidating like a kernel, the germ of the visible world arises out of the invisible. It is the visible reflection of the beings working through it. As the artist recognises himself in his work, in its success or failure, its progress and set-backs, so is it with the spiritual beings.

Out of their activity, out of their work, as it gradually draws to completion, there arises for them the experience of their own life and being; they behold themselves as in a mirror.

THE FIRST HUMAN GERMS

Such is the majestic sight open to the consciousness that is able to look upon it. Such a consciousness recognises as the first and lowest of the " Hierarchies," in embryo as yet, the very beings who, in time, are to become men. *They* are there from the start. A dwelling-place is made for them through the co-operation of their sublime creators. Itself, to begin with, still in the womb of its divine origin, the germ of this dwelling-place is first filled with cosmic life, is then ensouled, and after vast ages attains at last to the solidity and concreteness of the present day. Thus has the earth arisen, and as the nucleus of all that is visible, placed itself over against the invisible, parting from a universe which remains united to the spirit. Thus is it able to sustain those beings dedicated to a special task; the men of to-day. For this is the mystery of the earth's evolution: to be cut off from the divine worlds, and by that exclusion from its pristine state to find itself, and so again to interweave its proper activity into the whole. Henceforth, *earthly dwelling-place and cosmic home* are the two poles between which all evolution progressively revolves.

Long intervals are interposed, during which all that has " become " is taken back into the womb of the spiritual. The accomplished work returns to a state of becoming; reverts to the stage of being able to choose; then, harbouring what is chosen, the time ripens for a new manifestation. Three vast epochs of world-evolution, separated from one another by such " world nights," pass by before the ultimate germ-seed of the earth is set, before anything visible can emerge. All the Beings, present as the creators and witnesses of the earlier stages, again take part in the later ones; but they have risen, once, twice, three times, to ever greater perfection. Others have remained behind: their activity will find its place later in a changed form. At each stage, like a new principle something new is united with the earth's content. In the first stage the germ for what is later to be the physical body arises; in the second stage the life-body is fitted in to it; in the third, is added the vehicle of feeling, and of the inner life in general (the astral body): *only in the fourth stage does the human being as such enter into the " body" now sufficiently matured.* Only now does man become an individual. Now he can attain to self-consciousness, to the capacity for individual action.

The significance of the above terms should be studied in the writings which give the basis of Rudolf Steiner's work (1904, 1909, 1925) (Note 14).

THE KINGDOMS OF NATURE

But not all the germ-organisms originally destined to become men could keep pace with this evolution. At each stage a number remained behind, to re-appear in the next stage as somewhat imperfect forms. Thus, corresponding to these three stages, there is laid the foundation of the three kingdoms below man: animal, plant and mineral. *Only the human kingdom has advanced uniformly, and, at the earth stage, has attained a body able to be the bearer of an " ego."* In the other kingdoms the past lives on: in the mineral kingdom the first, in the plant kingdom the second, and in the animal the third world-epoch. But man too bears in the structure of his being the traces of the past; the physical body he has in common with the minerals, the life body with the plants, and the astral body with the animals, though all these three are permeated with his " ego," which raises him above the kingdoms of Nature. The gaps between the Nature kingdoms—which up to now have remained a riddle to scientific investigation—are the persistent echoes of separate world epochs.

We must refrain here from any description of the cosmic processes which accompany this coming into existence. Like the happenings on earth, they are the expression of the deeds of Spiritual Beings. The forming of the sun and of the moon in relation to the earth will now be considered.

The course of the four great world-cycles stirs us like the movements of a grand symphony. It differs from the customary picture of evolution chiefly in this, that long and ineffaceable intervals (" rest pauses ") mark off the various epochs one from the other, and that, always at the beginning of each new period, some quite new principle is at work. This new principle, seen from the aspect of the Spiritual Beings active in it, is described by Rudolf Steiner as the intervention of new hierarchies in the process of evolution. Looked at in relation to the gradual physical embodiment of earth and man, each new stage reveals itself as the addition of a new elemental condition. While the primal condition, the " old Saturn," in its substance is comparable only to the element of warmth, which for human experience lies on the borderland between spirit and soul on the one hand and the physical on the other, in the second world cycle (that of the " old Sun "), the air-like element is added; in the third (the " old Moon ") the watery element; and only in the fourth condition—which is that of the present day—is the solid element added. . . . For this very reason, this latest state, considered as a cosmic epoch, is also called " Earth." Hand in hand with the formation of these four elemental conditions goes the evolution of the four kinds of ether (etheric substance), a description of which lies outside the limits of this book. Closer knowledge of these four world cycles is indispensable for a real understanding of the processes in cosmos, earth and man (Note 15). That it is also decisive for the comprehension by Natural Science of the facts of evolution is now to be shown.

The following table summarises the description elaborated in detail by Rudolf Steiner in many different places (Note 16).

Old Saturn-evolution:	Old Sun-evolution:	Old Moon-evolution:	Earth-evolution:
			+Life Ether
		+Sound Ether———	—Sound Ether
	+Light Ether———	Light Ether———	—Light Ether
Warmth-Ether[1] or Element	——Warmth———	Warmth———	—Warmth
	+Air Element———	Air Element.———	—Air Element
		+Water Element———	—Water Element
			+Solidity

For our purpose it is particularly necessary to direct attention to the fourth epoch, that of Earth evolution proper. Here only are the foundations laid of that " world " which confronts the

[1] Note.—The state of warmth is not recognised in the physics of our day as an actual condition of densification.

scientific enquirer of to-day. This makes it possible for him to compare his own investigations with Rudolf Steiner's statements. Yet everything that is formed in the Earth epoch remains meaningless, " signifying nothing," if we ignore its spiritual aspect. *The detailed and complex " becoming " of the master work " Earth "* through the various stages of condensation from warmth, through air and water to earth throughout long epochs (Astronomers, too, speak of condensation for immeasurable periods of time), only acquires significance inasmuch as the mineral earth is destined to be the bearer of the human form to-day. The previous stages are all a preparation for that Earth-consciousness, which is imparted to man only upon that " earth." Only here does man attain the " Ego " as his personal possession. Hitherto this " Ego " still lived with the Creative Beings surrounding the Earth: now it descends and unites with the physical body of man, which is sufficiently matured. *Man becomes the bearer of the " Ego."*

THE GEOLOGICAL EPOCHS

The beginning of the earth's mineralisation is marked once again by a recapitulation of the epochs—a recapitulation, this time, *within the solid mineral element. Consequently, though each epoch is itself accessible only to spiritual investigation, yet echoes and reflections of both have been preserved in a form accessible to ordinary research.* Rudolf Steiner himself stated specifically that the periods named Azoic and Palæozoic by the geologists represent belated recapitulations of the far earlier Polarian and Hyperborean epochs. " Just as the seed of man develops in the embryo period . . . and in doing so recapitulates in little all the phases of mankind's development, so the earth at her birth into the mineral state recapitulates her own earlier spiritual stages " (E. Pfeiffer. 1926).

According to Rudolf Steiner's statements:

The " Polarian " Epoch is reflected in the geological period called	Azoic
The " Hyperborean " Epoch is reflected in the geological period called	Palæozoic
The " Lemurian " Epoch is reflected	

in the geological period called	Mesozoic
The Atlantean Epoch which follows, represents the geological period called	Tertiary
The Post-Atlantean Epoch represents the geological periods	Diluvium (glacial drift)
and	Alluvium

ATTEMPT AT COMPREHENSIVE SURVEY

If the effort is made to bring this description, of Rudolf Steiner's, of the descent of the different kingdoms and groups into connection with the geological eras given here, we arrive at the following results, which at the same time answer the questions asked by the palæontology of our time.

AZOIC ERA

During the Azoic era the solid mineral matter of the earth is separated off for the first time. But it can only be taken up by those beings who are most backward in evolution; hastening ahead of the others, they "embody" themselves in the new substantiality. Thus the minerals appear as the first solidified kingdom on the earth. The remaining kingdoms, though already existent in rich variety and even at this stage divided into plant and animal and differentiated into types, still resist solidification. Their archetypes have remained behind hitherto in the more plastic elementary kingdoms; now, as they begin little by little to incorporate the solid elements, they take form as those organisms which belong to the great order of invertebrates. These were previously *already* differentiated as to their essential being, and their separation into types was an accomplished fact before the mineralisation of their bodily form set in. And here we have the answer to one of the great enigmas of palæontology—the plasticity of the ancestral forms and the lack of true intermediate forms able to relate the chief types in the geological records. The ancestral forms did indeed exist in bodily shape, but they were of the finer substantiality which Rudolf Steiner has described as belonging to pre-earthly conditions, composed of the elements of warmth, air and water, *without as yet having absorbed any solid matter.* This supplies the reason for that mutability which has up to now completely baffled research. At the same time, we see why no impressions nor remains of them could be left in the earth's crust.

Thus the germs of the future human body are among those forms which resist densification.

PALÆOZOIC ERA

The Palæozoic era which follows is a world in which the plant kingdom predominates. In it the coal and slate formations prevail with their essentially plant-like fossils. Not until the Cambrian period do the animal groups (brachiopods and trilobites) take on a mineral body which reveals that the hardening process is intensifying. The first fishes (in the Silurian period) and the first amphibians (in the coal measures) are armoured like lower animals. There comes to light in the Palæozoic era an extraordinary range of animal forms: a clear indication that their differentiation had been accomplished before this era.

At this stage man still retains a plastic body, densified no further than the fluid stage. His psyche is still guided by Higher Beings. His consciousness is still plant-like, sleeping.

Another characteristic of this epoch is the dominance of a sense organ which has remained a problem for external scientific investigation; an organ which appears in the legends as the one eye in the forehead of the giant, and has consistently been identified with the pineal gland.

E. Dacqué, who by his own statement (1924, p. 349) allowed himself to be " stimulated " by the affirmations of occult literature, characterises the possession of armature and the median eye as the hall-mark of the Palæozoic times, and he ascribes the same organisation to the man of that period. If it is not forgotten that man at this period had not yet taken up the solid element into himself, such an idea can readily be accepted. It must also be remembered, however, that " Light " in those times was not yet the optical impression it is to-day, but rather an active agent penetrating the beings, forming in them the soul nature and the physical nature, as it does still for the plant world to-day.

MESOZOIC ERA

It is in the Mesozoic era that the special quality of the Lemurian epoch first comes to full expression, for the previous periods were still echoes of the Polarian and Hyperborean epochs. The Lemurian epoch is distinguished by the incorporation of the solid element. Rudolf Steiner describes how this solidification of the earth, as it progresses, becomes a danger which threatens the whole earth evolution with sclerosis and death.

If we wish to form a picture of the structure of man's ancestor at the beginning of the Lemurian epoch, we must think of the type of lower vertebrates (fishes and amphibians), but in forms less densified than their descendants of the present day.

THE SEPARATION OF THE MOON

In this time, therefore, occurs the great crisis of the earth evolution which leads on to the central event of the Lemurian epoch: the departure of the moon from the earth. This too is a cosmic event, and like the separation of the sun in the Hyperborean time, is the deed of Spiritual Beings. These Beings shield the earth from the definite hardening which threatens to overpower it, by separating from it all those substances which have become too dense and combining the latter into a new world-body which henceforth accompanies the earth as a satellite. *The moon represents the totality of the hardening dying substances: between it—the retarding element, and the sun whose development is in advance of earth's, the earth itself is now able to attain its further development.* It is at this point that the astronomical inter-relations of earth and its surrounding planets which hold good at the present time are contributed in their essentials. (However, the moon is more than a lifeless slag-heap. Lofty Beings, who influence later human evolution, remain connected with it.)

Through the separation of the moon, conditions on earth are completely altered. This event, which to all appearances is "only" a cosmic one, spells revolution for the life on earth. And this revolution must be treated with special attention in what follows: because through it, according to Rudolf Steiner, comes to pass the event which is *decisive for the ancestral form of man. The human form takes on the erect position.*

MAN'S ERECT FORM

Rudolf Steiner's power of vision throws a clear light at this point on the mystery of Man. Of the human form in the middle of the Lemurian time he writes (*Lucifer Gnosis*, No. 19-33, p. 45, first published 1905[1]):

"... It converted one half of itself with two out of its four organs of locomotion into the lower half of the body, whose chief function thereby became the nutritive and reproductive activities. The other half was directed upwards and the two remaining organs of locomotion were stretched out as fin-like hands. And those organs, which had previously aided nutrition and reproduction, began changing into organs of speech and thought. Man has raised himself erect. That is the immediate consequence of the separation of the Moon."

The upright position, which, as was shown in the first chapter, is characteristic of the human body alone, points to that cosmic event, the separation of the moon from the earth. But simultaneously and from the same cause, there takes place, according to Rudolf Steiner, the separation into two sexes.

Before this, man's ancestor combined the male and female elements, and reproduction took place without the participation of another being. Now that the human body continued to develop physically the organs of one sex only, and required to be fertilised through the agency of a second individual—" forces of the soul, with no outward task to perform, could unite with spiritual forces: and because of this union those organs were developed in the body which later made man into a thinking being. The power through which humanity forms for itself a thinking brain is the same through which in ancient times man fertilised himself. The capacity for thought is purchased at the price of sexual differentiation. . . . Male and female body, each presenting out-

[1] English Edition *Atlantis and Lemuria*, London (1911) and *Cosmic Memory. Pre-History of Earth and Man*, New York 1959.

wardly one incomplete aspect of the soul, are able for that very reason to become inwardly more perfect creatures."

The consciousness of man's ancestor had previously been dreamlike. Now, through the erect posture and the separation of the sexes, the ground was prepared for the consciousness proper to the earth period. Man awoke to an " outer " world.

THE APPEARANCE OF HUMAN CHARACTERISTICS

From now onwards, the head of the human form is directed to the cosmos, the limbs to the earth: the upper extremities can be transformed into hands, the lower into feet. Now too the speech organism can be formed (p. 17), whilst as a result of the separation of the sexes, the first germs of the thinking organisation are created (p. 18). Thus the foundation has been laid for the erect posture, for speech and for thought: those gifts which raise man above the other creatures of the earth, and enable him to grow out beyond the epoch in which he himself was still only a " creature " (Note 17).

It is significant that those vertebrates (fishes and amphibians), which are the creatures corresponding to-day to the earlier Lemurian period before the division into sexes, show in their anatomical structure even now traces of a bi-sexual disposition.

A proof, too, of the incisive significance of the departure of the moon is the fact that certain of the higher animals manifestly shared in the erection of the body, without however being able to attain it completely. For example, among the Mesozoic Theropoda, the Dinosaurs (tiger dragons), and more particularly the Ornithopoda (bird dragons), were able to carry their gigantic bodies erect on their strong hind legs, while some of the latter had even an air-containing skeleton like the birds. Their appearance geologically signifies that the separation of the moon is already an accomplished fact. This obliges us to place within the Mesozoic period that cosmic event which gave man his sign manual—the erect form. The extraordinary transformation of geological character, which took place between the Jurassic and the Triassic periods, leads us to judge that the actual time lay between these two.

This confirms the suppositions of the scientists mentioned on p. 44, that man's individual characteristics were expressed in his bodily form very much earlier than is usually conceded, i.e. as early as the Mesozoic era. Dacqué also reached the conclusion that the raising of the body to an erect position was the distinguishing feature of the Mesozoic era (1924, p. 66).

MAN'S BODY REMAINS PLASTIC

To comprehend the further evolution of man, we must first realise that the human ancestor of the Lemurian epoch was made of other stuff than his animal contemporaries. Rudolf Steiner says in this connection (*Lucifer Gnosis*, 19-23, p. 33[1]):

" Along with man there existed animals, which stood in their fashion at the same stage of evolution as he. According to present-day ideas these would be classed with reptiles. Apart from them there were the lower creatures of the animal world. But between man and animal there was an essential distinction. Because of his still plastic body, man could live only on those parts of the earth which had not yet passed over into the hardest material state. And in those regions there lived with him animal beings whose bodies were equally plastic with his own. In other places, however, there lived animals with bodies already densified, and which had already developed sex-differentiation and the senses. . . . *These could no longer develop progressively because their bodies had taken on the denser materiality too soon.* . . . Man could progress to a higher form because he had remained in those regions which were favourable to his constitution at that time. For this reason his body remained pliable and soft enough to enable him to mould in himself organs which could be impregnated by the spirit."

Time and time again on the long path of evolution it has been this quality of *remaining plastic* which has made steady advance possible for man. At the same time there separate from him—true branches on the great tree of Existence—one new form after another, forms which in the course of their descent become geologically discernible.

[1] *Atlantis and Lemuria*, and *Cosmic Pre-History*.

THE LOWER MAMMALS ARE SEPARATED OFF

After the separation of the moon, the lower mammals constitute the new type which diverges from the human stock then entering upon the corresponding stage of development: these are traceable geologically from the Jurassic period. Morphologically they do not derive directly from the reptiles of the previous period, but form a separate branch which diverges later than these latter from the parent human stock. This is why no links are to be found between reptiles and mammals in the earth's crust. The late Lemurian period then is one in which all animals—the higher mammals, and especially the quadrupeds, excepted—are already in existence. The highest forms were no doubt those of the kangaroo group, the aplacentalia, distinguished to this day from other mammals by the much slighter connection between mother and offspring.

END OF THE LEMURIAN PERIOD

The Lemurian epoch came to an end, according to Rudolf Steiner, in a catastrophic series of events. Progressive " demonisation " brought about an unbridled overgrowth of formative forces, such as could not fail to bring destruction in its train; this destruction, which in *Geheimwissenschaft im Umriss* (1920, p. 263)[1] is described as a catastrophe of fire, was originally caused by the intervention of a spiritual influence necessary for evolution and which Rudolf Steiner calls " Luciferic." This Luciferic influence became excessive in certain regions towards the end of the Lemurian epoch with the result that the organic forms then living were distorted—became monstrous, fantastic, morbid—and accordingly bodies appear which seem to spring from a world of torturing dreams and hallucinations. Even to-day the geological remains of the animal world of that time shed a clear light on the crises of that epoch; the grotesque and awful forms of the saurians can still make the beholder tremble if he imagines them called back to life. Powerful and rapacious aquatic creatures, half fish, half crocodile; land animals that are moving mountains, half elephant, half kangaroo; giant spectres of the air, half lizard, half bat;

[1] English Edition *An Outline of Occult Science* (1922), p. 233.

all bear witness to a time fallen irresistibly under the sway of demonic power. It is as if the undisciplined soul world, not yet irradiated by an Ego, were once more gathering together all its baleful powers before the cosmic hour has struck for that earthly creature who is able, not only to carry these same powers within himself, but also to dominate them.

THE ATLANTEAN EPOCH

Not until the fourth epoch of the earth's development, the Atlantean,[1] does this new being arrive. It is in this age, at last, after long preparation, that the peculiar task of earth is accomplished: the human Ego can now take possession of the erect body. It has laboured long upon the physical organisation until the latter can become an Ego-bearer. Now man treads the solid earth with his feet, now he forms articulate speech, now from a dream-like consciousness a perception of an outer world arises for him, a perception still indeed permeated by clairvoyant vision. But a group of creatures has hastened to descend before him, to attain solidity before him: these are the *great land quadrupeds*, which reach their prime in the Atlantean epoch, corresponding to the Tertiary era.

THE GREAT QUADRUPEDS

Before reaching the stage of bodily maturity necessary for the reception of an Ego, they have separated from the human stock, and subordinated themselves entirely to the earth forces. They therefore develop further only in one-sided directions (see p. 19). The erect posture which they originally shared is lost to them. Just as fluid purifies itself by sedimentation, so the human parent stock, by throwing off those forms, frees itself from the last imperfections which stood in the way of its penetration by the Ego. Their remains are found in the early Tertiary (Eocene) era, viz., Condylarthra (forerunners of the hoofed animals), Credontia (forerunners of the carnivora), etc. From these were developed the higher mammals, which are still to-day man's neighbours, and

[1] Its spatial centre lay in the area that is filled at the present time by the Atlantic Ocean.

which have become still more dissimilar to the human type than were those early forms, which both in their limbs and their dentition were more primitive, that is, in reality, more human.[1]

SEPARATION OF THE APES

Again in the Atlantean epoch a group remains behind, which moves at the eleventh hour as it were to areas where it cannot save itself from solidification. From this group come the primates or highest order of mammals, in particular the apes. These are at the same time the last order of animals which can have a group existence only. All those who divide off later than these from the human stock, remain in the human kingdom and become bearers of an " Ego." The apes, however, prematurely absorb the forces of gravity into their bodies, cannot withdraw the head from the impression of weight, and bear from now on the stamp of the brute: this is the more noticeable the more certain of their characteristics remind us of that lost human relationship (cf. following p. 78).

MAN

Man alone continues to abide in those parts of the earth in which he can still remain plastic, until at last for him too the hour has come. He makes his geological appearance in the later Atlantean epoch (glacial drifts), and has, as we saw before, from the beginning, all the essential human characteristics together with signs of a developed culture. The fossil remains which the palæontologist finds to-day are not indeed those of men of the highest races (who carried on the Atlantean civilisation), they belong to side branches; even Neanderthal man, as Rudolf Steiner has expressly shown, is not a direct ancestor of the civilised humanity of to-day. Years after Rudolf Steiner, contemporary research has arrived at the same conclusion, and has placed Neanderthal man on a side branch of the genealogical tree (see p. 35 and Diagram 4). The culture and life of the Atlantean, even his speech and his manners, have been described by Rudolf Steiner

[1] The placentalia thus shared this evolution until the early Atlantean time. The placenta has a clear relationship to the reception of the Ego.

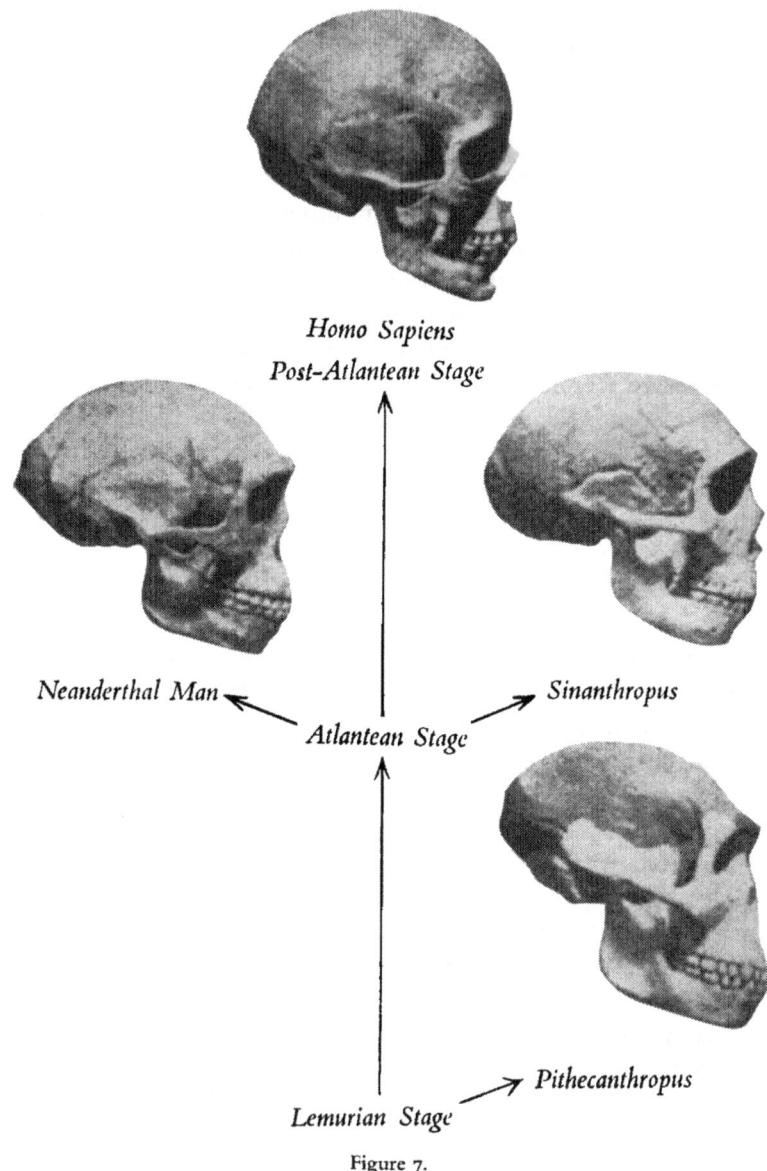

Figure 7.

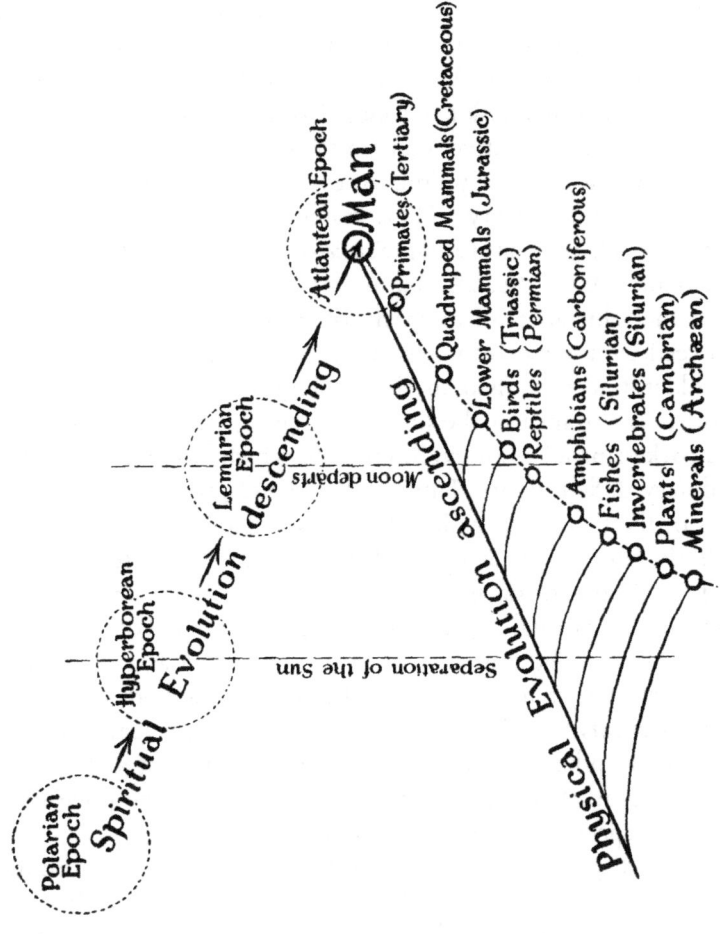

Diagram 8.—The spiritual and the physical aspects of evolution.

in his book *Atlantis and Lemuria*. In this book he depicts the wonderfully high culture of the true " primeval man " of which Natural Science, with its " cavemen," the contemporaries of mammoth and cave bear, knows only the degenerate descendants.

With the Atlantean epoch begins the history and pre-history of man. This was the time, first of the oracle and, later in the post-Atlantean epoch, of the mystery centres; from which all civilisation comes forth. The post-Atlantean sub-epochs, of which ours is the fifth, have it as their task to develop, from the original clairvoyance of man, clear objective consciousness, and from the tribally determined will of early times to evolve the free activity of self-conscious individuality. (This statement should be compared with the chapter in *Occult Science*—" The Evolution of the Earth and Man." See also Figure 7 and Diagram 8.)

In the reluctance, so to speak, of Atlantean (Tertiary) man to proceed to mineralisation, we perceive the secret of the absence of older traces of man in the geological record, and of their sudden appearance in the Diluvial (Glacial) periods. This solves the darkest problem of evolution, one which has presented extraordinary difficulties to Natural Science and provided the opponents of the theory of evolution with a welcome argument.

In the later Atlantean epoch then, when the descending spiritual and soul powers of man united with the ascending physical vehicle, the spiritual and the physical processes of evolution come into contact. The human " I " is now received as a kernel into the matured sheath of the body. Before this the bodily development was only a kind of shadow of the soul and spiritual kingdom; now the two halves have united to form a single process.

In Diagram 8 an attempt is made to illustrate the polarity of the two aspects and their final union.

In this diagram the spiritual and the physical aspects of human evolution are shown as branching off from one another. It might therefore be misinterpreted as suggesting an evolutionary development in the direction of dualism. In fact, however, the two diverging lines merely represent the same process seen from two different points of view, the spiritual and the physical.

Admittedly it is not easy for a modern intellect, first to observe the diagram and then to fit the two lines together and grasp them as one—in a final act, as it were, of thought. But the fact that precisely this fusion is *contradicted* by the diagram arises from an imperfection inherent in diagrammatic representation.

The lower line shows the natural " stages," that is, the series of sense-perceptible facts actually available to scientific investigation of Nature. In order to present the total spiritual-physical process, it is not enough just to add to these a series of abstractly conceived " spiritual impulses." That *would* be " dualism." The phenomenal series must rather be completed by showing the series of equivalent stages of spiritual evolution. These, it is true, are *perceptible* only to spiritual vision, but their complementarity can be intellectually grasped. They are the actual beings themselves, whose sheaths develop step by step along with them and undergo changes in the process.

The act of union which the diagram follows out is, thus, not simply the sinking down of these beings into a bodily element which has been growing up from another direction; it is their final fusion with a physical counterpart which has all along been progressively adapting itself for the purpose. Every step in spiritual evolution *is* at the same time a step forward in the birth and growth of the " sheath."

Unsatisfactory as it no doubt is, the " two branch " scheme of representation has this advantage; it shows how the spiritual stages described by Anthroposophy coalesce with the evolutionary stages gone through by these " sheaths "—which are described by natural science—so as to form one single intelligible process.

RECAPITULATION

When we survey the picture of " the descent of man " which this new theory of evolution puts before us, we recognise that man's ancestors provide the original root-stock from which all other forms spring. The forms of the lower, higher and highest animals (invertebrates, vertebrates, amniota, warm-blooded animals, placentalia, primates) have departed from the up-growing trunk of humanity as side branches, and at points which come ever higher up the tree. But as far as bodily perfection is concerned, the higher mammals have passed beyond the human form into restricted evolutionary blind alleys.

But the chain of creatures leading up to the being and becoming of man runs " like a golden thread through the winding web of things created." *The ancestry of man is the inner bond that holds evolution together.* The enigmas of the genealogical tree are solved by Rudolf Steiner's spiritual picture of evolution.

A NEW EMBRYOLOGY

Seeing the theory of descent in this new perspective, however, involves far-reaching readjustments of a second vast group of facts—embryology. So far we have avoided this subject. The time has now come to deal with it.

Today, when Haeckel's " biogenetic law " is the common property of educated people, there is no need to speak of it in detail. As early as 1811 Meckel had noticed the " similarity " between the stages of development of the embryo and the types of the animal kingdoms. Later, Fritz Müller (1864) gave a definite direction to embryology by indicating its relation to the theory of descent. But it was through Haeckel alone (1866) that this " biogenetic law " became common knowledge: in the arresting formula, " the development of the embryo is a condensed recapitulation of the development of the species." It is not too much to say that, thanks to this comprehensive view, embryonic development was changed from a " waste heap of unwieldy raw-material " into an intelligent process. An unintelligible game of Nature's own in which she appeared to delight in following the strangest round-about ways, all at once became a clear legible record of the past.

The facts spoke clearly enough. The larva of the frog (tadpole), for example, has a rudder-tail, breathes through gills and has a heart consisting simply of an ante-chamber and a main chamber; when referred to their fish-like ancestors, these peculiarities were comprehensible. The embryo of the higher vertebrates and of man has pharyngeal slits: the interpretation of this fact as an evidence of their aquatic ancestry could not fail greatly to impress the generation which discovered it. The earliest ontogenetic stages of all multicellular animals include a gastrula stage; it was possible to construe this as a striking proof of common descent from an original cup-shaped larva (gastræa). With an ever increasing number of such parallels embryology became such a powerful support to the theory of evolution that it was classed by Haeckel as the third "science of origins" side by side with palæontology and comparative anatomy.

But this Haeckelian interpretation of the new law presented difficulties. K. E. von Baer had already urged caution. The embryo, he said, does not pass through the forms of fully developed animals, but repeats only certain details of their organisation. A human embryo is never actually similar to a fish, it is only that its pharyngeal slits remind us of the gills, and its flattened limb construction of the fins of fishes. Moreover, the embryo at no stage of existence passes through the phases of development of any other main type (phylum), but decides at the beginning and quite irrevocably on a definite type; and then specialises more and more, so that the whole growth of the germ is a self-determination towards the *differentiæ* of class, order, family, genus and species, accompanied by an ever clearer distinction from similar forms. But in spite of this objection, it was felt that there must be some truth in the "rule of recapitulation"; for rudimentary organs, such, for example, as the pineal gland in man, by the manner of their original disposition in the embryo, point to some original purpose (the frontal eye) which they no longer fulfil, so that wherever development obviously enters upon by-paths, an indication of the past had to be recognised.

Haeckel, having recognised that the correspondence between the development of the embryo and the evolution of the race was incomplete, brought forward the theory of "disturbed development" (Cenogenesis) as a corrective to his law. The embryo, he argued, could not of course repeat all the details of race evolution; it must shorten this, missing out certain stages; moreover, because of the indispensable auxiliary organs and appendages, such as the yolk-sac, and of sheaths like the amnion, and yet more through such transitory organs as, for example, the tadpole's beak, the image of many of the stages is obliterated and even falsified. We had to learn to read the record aright in spite of its gaps and disfigurements.

But behind this reservation lurks an obvious danger; it was liable to abuse in that, starting off with a ready-made picture of the history of evolution, men now read into the history of the embryo the form they were looking for, and everything that clashed with this, they accounted for by Cenogenesis. It was no longer of the slightest importance that the embryo never resembles a fully developed animal; however weird its appearance, still it could always be said that it repeated the ancestral stages, since, thanks to Cenogenesis, one part of the historical representation was, as it were, treated as a " fable convenue."

In the early days of the evolution theory it was easier to disregard this difficult position than it became later on. But even so it has become customary to describe all embryonic appendages and all sheaths as adaptations to life in the egg and in the uterus, and, as special formations, to exclude them from any phylogenetic consideration. Meanwhile the forms of the embryo itself continued to be regarded, as at first, as repetitions—imperfect indeed, yet of use to the investigator—of a sequence of ancestors, and this seemed to be confirmed by comparative anatomy and palæontology.

Now, however, through Rudolf Steiner's theory of spiritual evolution, embryology is placed in an entirely new position, for if, on the basis of his statements, we have to admit that the progenitors of man and the animals lived under earth conditions completely different from those of the present; moreover that the former were not comparable either in shape or in substance to the creatures on earth at this present time—then *at once these singular embryo forms take on a new and unexpected interest*. The very appendages and sheaths have suddenly become significant.

CORRECTION OF THE THEORY OF ORIGINS NO LONGER NECESSARY

All allowances for alleged ontogenetic distortions of history cease to be necessary.

The text can be studied with absolute fidelity now that it is no longer necessary to be on one's guard against these distortions; and a wonderful task is opened up to embryology when it is able to consider the stages of ontogenetic development, as they actually *are*, as the images of ancestral forms. These ancestors were not creatures of earth like those of to-day; so that when studying the embryonic stages, we are looking in truth into the preceding world-stage of the earth. That world, which Rudolf Steiner by means of super-sensible vision describes in its genesis—that world finds its reflection both as to form and as to content in the forms through which the developing germ has to pass. At last the strangeness of the early stages becomes comprehensible; and now it is just those which in the last century were thought distorted which emerge as the most valuable evidence of the misunderstood past. Only certain preliminary indications can be given here as to what course this new reading of the historical record may take: there is work in it for decades.

Immediately before the first edition of this book was printed two important treatises on human embryonic development were published by Karl König (1927). His conclusions are of exceptional importance for the problem—" man and animal," but unfortunately no more than a brief summary can be given here.

König proves convincingly that the essential characteristic of the growth of the human embryo is just this, that *the formation of the embryo takes place relatively very late*, the first weeks being devoted to the formation of the embryo sheaths and appendages (yolk-sac, allantois). *Thus the embryo is formed by its surroundings, just as in cosmic evolution the human form is the work of the cosmic forces surrounding the earth.* These forces stream down one after the other at the different stages (which König describes), just as the etheric powers—warmth-ether, light-ether, sound-ether and life-ether—did at the beginning of the earth (cf. the plan p. 51). Starting from the original form (morula) the first event that takes place is the separation into *Trophoblast* (outer) and *Embryoblast* (inner), then the latter separates once more into *Magma reticulare* on the one hand and the true embryonic nucleus (wholly internal) on the other, and König rightly sees in this a recapitulation of the first three epochs of the earth's evolution:

	POLARIAN EPOCH:	HYPERBOREAN EPOCH:	LEMURIAN EPOCH:
Microcosmic:	Undivided morula	Trophoblast +Embryoblast	Trophoblast +Magma reticulare +Embryonic nucleus
Macrocosmic:	Undivided primal body	Sun +Moon-Earth	Sun +Moon +Earth

Only in the third period (which corresponds to the Lemurian) does the embryonic cell divide into two small vesicles which are the foundation for amnion and yolk-sac. The yolk-sac even detaches from itself the yolk-sac vesicle, which König interprets as analogous to the moon's separation from earth! Only *after* the allantois structure has emerged from the yolk-sac does the true embryo " take." It forms last, after all the " sheaths " are finished.

What is distinctively new in these researches is that here for the first time the sheaths and appendages are given their full weight in the deciphering of the embryological record. It becomes manifest that the egg sheaths, the allantois and the yolk-sac themselves, contain the most significant indications of the past history of the cosmos. Far from being " merely " auxiliary organs, their formation reflects the cosmic stages of the earth evolution, its Polarian, Hyperborean and Lemurian epochs.

NEW FORM OF THE BIOGENETIC LAW

In agreement with Karl König's researches we can deduce the future formula of the much contested biogenetic law:

The development of the human embryonic sheaths and appendages is a reproduction of cosmic events in a material medium.

How does it stand with the development of the embryo itself? This only begins *after* the formation of the sheaths and appendages. What then, we ask, is the significance, as a record, of the successive life stages of the developing human form?

It is a truly sublime thought that the successive embryonic stages of the human body repeat, in physical form, the events of the last epochs of spiritual evolution, as described by Rudolf Steiner. In what follows, the author proposes to indicate the results of his own examination of the embryo, in the hope that the new embryology will devote itself to a more exact study of the correspondence in question.

Consider, to begin with, the delicate form of the human structure in its early stages; the form in which it first attracts notice is that of a longish streak, with definite inner articulation, at the place where amnion and yolk-sac lie one on the other (Haeckel's so-called " Sandal-germ "); how totally different it is in its appearance from the grotesquely swollen form of the following weeks, say at the end of the second month; and then in the third month how suddenly there arises out of this the form (human, albeit still strangely unearthly) of the fœtus. We have only to add the earliest stage of all, when the germ is still undifferentiated within the perfected sheaths (third week), and it is easy to recognise, *in these four principal stages of the embryo, the reflection of the four stages in the evolution of the being of man* (cf. with the following Fig. 9 on p. 72).

The first trace of an embryonic form, in the act of gathering itself together from the not yet differentiated or folded germ sheaths (A), reflects the first period of a purely physical stage of the pre-historic human ancestor. Here the soul-spiritual being of man remains as yet wholly outside the diminutive form. According to Rudolf Steiner the soul-spiritual being of man unites at the end of the third week with the germ, which up to that point can be considered only as a physical structure. When the tendency to take on the sole-shape (B) is developed, and the first segmentation begins to appear in the primitive spinal column, we can perceive in the now plant-like structure the *life-body* taking possession, and our thoughts will readily turn to the geological prototype of this stage in that (palæozoic) period which is indeed characterised by the preponderance of plant life. Here the human germ still grows like the plant in the sunlight, the inner organs are not yet distinct and separate; they are still distributed among the germinal " leaves," the segments are reminiscent of plant stems; the etheric or life-body works in them. Then, more clearly defined, the self-contained form arises, the swollen incurved semicircle, with organs sprouting internally and externally: (C) an image of

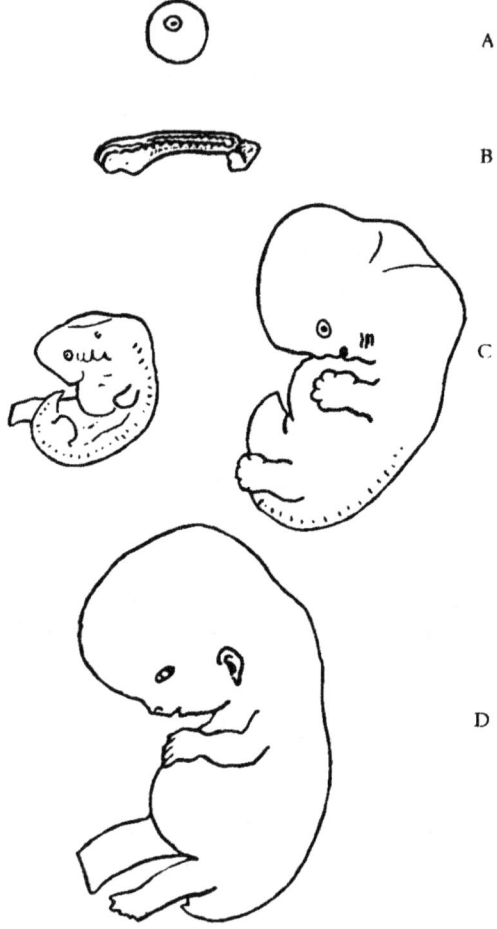

Figure 9.
THE REFLECTION OF THE FOUR EARTH EPOCHS IN THE LIFE HISTORY OF THE HUMAN EMBRYO

Top (A) human egg; immediately below (B), sandal germ (three weeks); next below (C), two embryos of five to seven weeks old; bottom (D), fœtus of the third month (after His).

the time when the *astral body* takes possession. Now the first sense organs become visible, the large eye, fixed as yet, and the little nose depression; on what is to be the future neck are the problematic pharyngeal slits, while the limb extremities, like flattened buds, await their future development: directly beneath the front end, which is rolled inwards, rises the heart in an immensely swollen arch. Once we realise from Rudolf Steiner's description that the organisation of the astral body stood wholly under the sign of the form-giving forces of the world of sound (cf. p. 53), we begin to understand the significance of the pharyngeal slits in the embryo: they are not reminiscences of the gills of fish, but are a recapitulation of the organ through which the human ancestor of the Lemurian epoch received into himself the form-giving vibrations of the resounding cosmos: they are a memory of the time when the human form was wholly a being receptive of the resounding cosmic tones. (As a matter of fact, considered in its entirety the embryo does still remind us of the outer part of the human ear.) Rudolf Steiner also explains how in those creatures which found their way into the watery element, this organ was metamorphosed into the gill apparatus for taking in and giving out water; while other creatures adapting themselves to air and earth, formed out of it the organs of hearing and of utterance. But both the speech organisation and the gill apparatus, recognised by comparative anatomy as being very closely connected, point back in their original form to the Lemurian epoch, and through this again to the Old Moon period, of which this epoch was a repetition. That was the time when the astral body entered man, that body of which Rudolf Steiner said that it was formed entirely out of tones.

In the third month of foetal life the head begins to enlarge enormously, the curvature of the neck is straightened, and the trunk stretches out (D). There can be no doubt that in this process there lies a reminiscence of the rise to the erect posture—the event which is connected with the separation of the moon (cf. p. 56). The perfecting of the limbs into hands and feet is next accomplished—that is to say, the polarity of upper and lower becomes visible. The human ancestor, whose reflection this stage is, approaches his union with the earth forces. Gradually the form of the embryo is stamped with the seal of the Ego, and this has the effect of restraining the astral body, which up till now has dominated the form.[1] The ossification of the hitherto cartilaginous skeleton, is the counterpart of the mineralisation of the earth. The development of the last months of fœtal life indicates the early Atlantean epoch. In the middle of the Atlantean epoch man stepped forth into the world: this may be thought of as reflected in the process of birth. Yet the consciousness of the new-born babe and of the young child is just as little open to this present sense-world as was that of Atlantean man, whose inner being felt for a long time the after-effects of a gradually fading clairvoyance. Only in the third year of life does the child reach the point when it learns to say " I " to itself and can bring itself as " I " into contrast with the world of things (see p. 119). Thus it is only at about this period that the descending Ego actually takes up its dwelling in the earthly body, and that point is reached where the descending spiritual evolution and the ascending physical evolution meet one another (Note 18).

The development of man—including the sheaths and embryonic appendages—as well as that of the child to the third year, is thus truly a repetition of the past, only in a far more extensive and exact sense than was dreamed of by Haeckel with his biogenetic law. Only through Rudolf Steiner has *the extension and deepening of the relation discovered by Haeckel become possible*. Perhaps the most general statement of what has taken place in man may be expressed in these words: *Microcosmogony is a reflection of the macrocosmogony.*

★ ★ ★

RETARDATION AS A PHYLOGENETIC PHENOMENON

At the close of this chapter attention may again be directed to the main theme of the present book, namely the essential difference between man and animal. The consideration of actual evolution has resulted in new light being thrown on this essential difference. The morphological riddle of the evolution of man is solved

[1] See page 80.

through the teaching of spiritual evolution. *The phenomenon of retardation takes on the aspect of racial history.* For the whole of evolution becomes an alternation between self-restraint and further development.

On the spiritual side of evolution taken as a whole, the difference between man and the other kingdoms arises because man alone keeps step in the march of evolution, that is to say he alone moves forward at an even pace, while the other beings, in varying degrees, are left behind. If the seer's eye is directed to the oldest period of earth's past, Old Saturn, it encounters beings still imperfect who represent the ancestors of *all* the kingdoms of Nature—alike of men, animals, plants and minerals. These are at the stage of consciousness of the present mineral kingdom (deep, trance-like sleep); it is nevertheless justifiable to name these beings human germs, since all of them in that first epoch are potentially capable of striving upwards to the human stage. Further evolution consists, as previously indicated, in the perfecting of the human germs; it is an unfolding of the possibilities latent within them: on the Old Sun through the reception of an ether body: on the Old Moon through penetration by the astral body: while in the Earth-epoch proper the " Ego " enters on the scene. At each of these stages, as already described, some of the human germs remain behind, having failed to take with their fellows the succeeding step in evolution; and it is in this way that the gaps between the kingdoms of Nature arise. The present-day natural kingdoms below man are a reminiscence of partings which occurred long ago during the course of general evolution. Thus the group souls of the present-day animals were themselves formerly on a par with the human egos which have now become individual souls—they were endowed with the same inner possibilities. They did not share in the transition to individualisation and have accordingly lagged behind, while man has gone forward. The stream of soul and spiritual evolution which leads up to man has left behind it the other kingdoms of Nature, including that of the animals, at the various stages of its progress.

From the purely physical point of view evolution takes on

quite a different aspect. So as not to complicate the picture unnecessarily, let us consider only the earth epoch proper, from the Polarian period onward. The development of the physical images of the already discrete kingdoms and types corresponds with the genesis and first appearance of the various elementary conditions. And here, in reverse order to the evolution of soul and spirit, that group of creatures makes the greatest advance which does not unite with any newly appearing element, and which therefore does not admit the new element into its physical form. This abstention undoubtedly depends on the inner disparity of the creatures (the question, how this disparity arose, is, if possible, as absurd as the question—what happened before Time came into existence), but the essential fact remains, that in evolution, considered from the physical aspect, there arises the distinction between bodies which are more capable of perfection, and which still remain plastic, and those which become less capable of perfection and gradually stiffen. Now some of the soul-beings (entelechies) are able to continue the ascent, by selecting for themselves the plastic forms—while the others unite with those forms which are losing their plasticity. This difference becomes particularly critical in the time shortly before the separation of the moon in the Lemurian epoch: at that time only a few isolated human souls are able to remain behind on the earth, while the rest seek out other scenes of action (cf. *Occult Science*, 1923, p. 214). Their " stiffening " bodies were left to less perfectly developed beings, which became the ancestors of our present animals.

THE SPIRITUAL BACKGROUND OF THE SEPARATION OF NEW FORMS

It is a remarkable process; certain beings leave behind them the result of their work, i.e. fully formed bodies; and other beings, below them in development, take these over. In this way the spiritual reality which lies behind all the ramifications can be realised, for it may be taken for granted that such partings as those described above must have occurred over and over again. The animal world, as it has come down to us to-day, and more particularly the higher animals, form what may be called side streams, which have left the main stream and are now, as it were, drying up. The representation of these side branches and of their surrender to the " hardening process " is a task for the zoology of the future. In this connection interest will properly be directed to the way in which the separate groups adapt themselves to particular elements, taking within these their characteristic form. Rudolf Steiner outlined in a magnificent way, in his Autumn lectures given in Dornach in 1923, the basis of this new zoology.[1]

The embryology of the future must attempt to show in some way in its tables of comparative embryology, on the one hand, how the human parent stock stops behind and remains plastic, and on the other hand, the overhasty progress and stiffening of the animal stocks.

One way of doing this is to place the subjects side by side, as indicated in Figure 10. Here the attempt has been made to

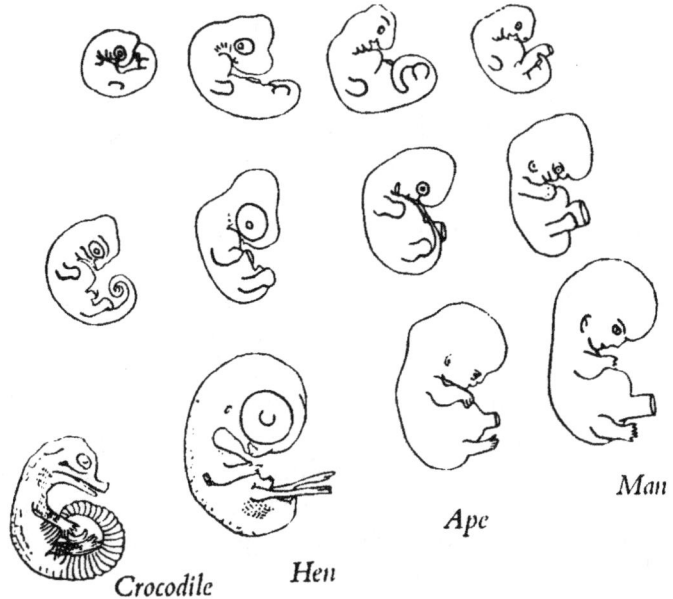

Figure 10.—Embryonal stages of the Crocodile, Hen, Ape and Man compared, illustrating the law of retardation.

[1] Der Mensch als Zusammenklang des schaffenden, bildenen und gestaltenden Weltenwortes. (English translation: *Man as Symphony of the Creative Word*. London, 1945.)

plot the phenomena of retardation graphically by showing the different degree of variation from the original form which the subjects respectively undergo at each stage of their development. Thus it is possible to illustrate the way in which the human form owes its universality to the circumstance of its remaining near its point of origin; while the animals become more definitely specialised and more unlike man the further and deeper they " descend."

COMPARATIVE ANATOMY

Comparative anatomy, in contrasts of the kind illustrated in Figure 11, is able to demonstrate clearly the varied working of the formative powers on the organs; it can for example show that the human skull remains dominated by the spherical tendencies which ray in from the surrounding world; while the animal skulls (ape, cat, dog, ox) become deformed in various ways, as the forces of gravity get the upper hand.

Anthropology too will be able in future to escape from its abstract statistical scheme of skull and body measurements, and to take part in the working out of a new physiognomics of the human body. Such a study can for instance compare those human skulls which have the facial region more violently developed, the eyebrow ridges more prominent, the forehead flatter, the sutures less marked, with animal skulls which show the full development of what in the human skull are indicated merely as restrained tendencies. Similar results could be obtained from a comparison of older with younger skulls (Virchow, as far back as 1870, noticed the likeness between the aged human skull and the skull of an ape, though he was unable to explain it).

The development of ape and of man after birth deserves special attention. The contrast between the two can produce an impression which amounts to a feeling of tragedy; the form of the ape, developed almost far enough to become a bearer of the Ego, nevertheless falls short at the very last moment and becomes distorted. It is indeed just that " almost " which makes the gap between ape and human being all the more palpable: there is a similarity both of form and of behaviour which amazes normal human beings—moves coarser natures to laughter, and pains the

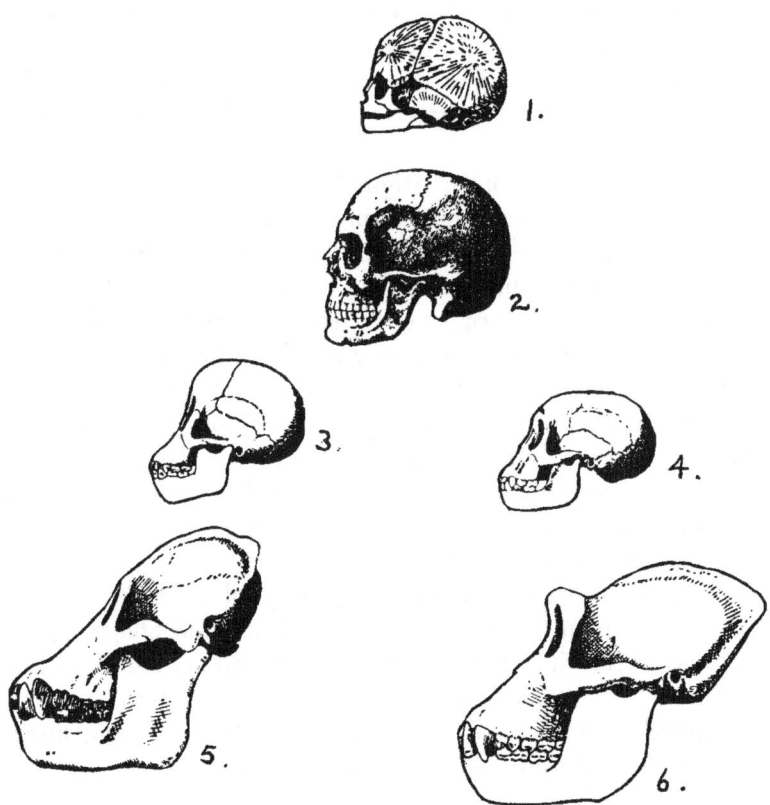

Figure 11.—Illustrating the degeneracy of the animal skull. (The animal skulls are reproduced with the kind permission of the publishers: Julius Springer in Berlin, from the work of J. Naef in the journal *Die Naturwissenschaften*, fourteenth year, Nos. 20 and 21, 1926.)

(1) Skull of a human fœtus immediately before birth; (2) a human adult; (3) A young orang; (4) a young gorilla; (5) an old orang; (6) an old gorilla.

more sensitive. Skull and skeleton receive too soon into the upper region of the body the stamp of the forces of the earth; the lower part is not yet adapted to bear the weight.

EARTH FORCES AND THE EGO ORGANISATION

It is here that the varied role of the earth forces becomes visible. These help the human being who is ready and mature to receive them towards the inner strengthening of that soul-spiritual organisation which is to become the organ of the Ego; whereas upon the apes, which take these forces into themselves at an immature stage, their effect is to harden and stiffen the bodily organisation in such a way as to shut it off from penetration by the Ego.

The ape, in spite of his original resemblance to man, grows later on into the very physical image of astral body not penetrated by Ego, while man on the other hand visibly impresses the Ego into his bodily form. " In the astral body the formation of the animal has its origin; outwardly the form as a whole, inwardly the formation of the organs. . . . Where this process of formation is carried to its conclusion, the animal nature is produced. In man, it is not carried to its conclusion . . . it is drawn into the realm of a still further organisation, which we call the ' organisation of the Ego.' . . . Down to the smallest particle of his substance, man in his form and configuration is a product of the organisation of the Ego."[1]

This gives us the key to the morphological difference between man and animal (see the chapter on " Form and Shape "); and at the same time it affords the simplest explanation of the divergent evolution of the two kingdoms: to become animal is to have been completely shaped by the astral body: to become man, means that the Ego has imprinted itself throughout the form.

FRESH GROUNDS FOR THE SCIENCE OF MAN

This fundamental process of the imprinting of the Ego on the human organisation opens up quite new territory in the science of man, to which the natural scientist, the doctor, the teacher and the artist can devote themselves in a common effort. A few indications only can be given here. Penetration by the Ego presupposes that the organism remains undetermined even *after* the age of early childhood. Only in this way can the form gradually become the image of individuality. The Ego, which until the third year worked from without inwards, is able subsequently

---

[1] Grundlegendes zu einer Erweiterung der Heilkunst von R. Steiner und I. Wegman, 1925, Seite 28-29. (English translation: *Fundamentals of Therapy*, London 1925 p. 29 et seq.)

to work, as it were, from within outwards, right from the centre to the surface; and this is the way in which man reaches physical maturity.

That the human organisation remains open, while the animal organisation closes itself, to the penetration of the Ego, can be proved in a very striking way. Rudolf Steiner often spoke of two significant turning-points in the life of childhood and youth—the change of teeth and puberty: the former takes place about the seventh year, and the latter towards the fourteenth. Both these processes take quite a different course in man and in the higher animal.

The cutting of the milk teeth begins in man a few months after birth, and finishes usually with the second year. Then, for man, a very singular pause begins which lasts several years: and not until the seventh year is there again a marked activity in the region of the teeth.

The increase in the number of the teeth and the change of the teeth continue until the tenth year. As Bolk has emphasised, an exactly similar interruption is experienced in the development of the human reproductive glands: the growth and the differentiation of texture in the ovary is already completed by the fourth to fifth year of the child's life, but function begins only towards the fourteenth year, in southern lands rather earlier: thus a rest interval of very nearly ten years intervenes. Now when we consider what decisive events take place for the human being during just these two life periods, this remarkable postponement of bodily maturity must be recognised as the real foundation for the formative activity of the " I." *The postponing of physical maturity leaves room for penetration by the Ego*—of which the completed organism is to be an image.

This is quite otherwise in the animal, especially in the mammal. Here the two processes go on much more quickly and without pauses; indeed they overlap: already at birth some milk teeth are generally present (this is the case with apes: Bolk); and directly the milk teeth are complete the increase in the number of the teeth and the second dentition begin. The interval of several years which is so characteristic of man is nowhere to be

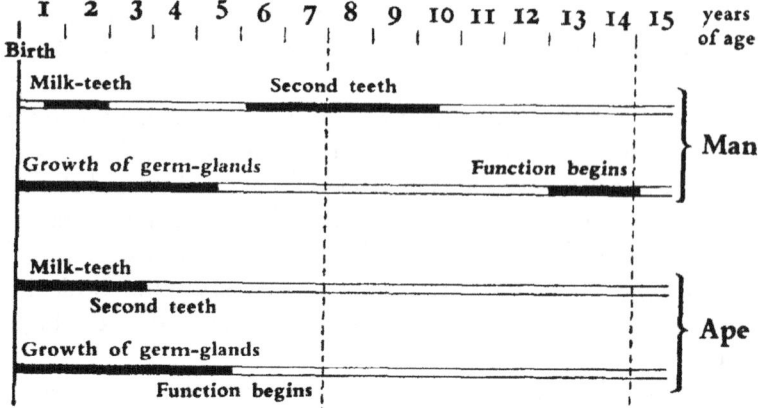

Diagram 12.—Table of the rhythm of development in man and ape.

found. The growth of the reproductive glands in ape and man is quite in agreement with this; their growth and differentiation of texture are perfected, as in man, in the fifth year of life; but then for animals, function begins at once; the age of puberty has come, for which, it need hardly be said, man has to wait for nearly another ten years (see Diagram 12). The physical development of the animal advances without a pause; no room is left for the imprint of the Ego.[1]

★ ★ ★

In conclusion, if after having grasped the full bearing of this difference, we look again at the picture of the two chimpanzees (Figure 1, p. 8), a feeling approaching to tragedy overcomes us anew, as we face this being which most resembles man: now, however, this feeling is heightened by our knowledge of what has taken place in the past, of which this striking picture is the record. We cannot remain unmoved before the picture of the young animal, with the face not protruding, the dome-like brow and head; the poise of the neck and (for its age) the full

---

[1] Treated at length in *Gäa Sophia* (Year Book of the Natural Science Section at the Goetheanum), Vol. 2, 1927, p. 237. Almost two decades later A. Portmann adopted the idea in his *Biological Fragments Towards a Doctrine of Man* (1944) and illustrated it with new material.

chest and human attitude of the forearm. We can positively *see* how the moment for becoming human has just been missed. And the accompanying picture of the full-grown beast reveals the completeness of the downfall; instead of the human profile, there is a pronounced animal head; and as for the dome-like cranium and brow, the erect carriage of head and trunk—gone!

The animal's inability to bide its time stands revealed before our eyes.

Another impressive example is to be found in the picture of an almost fully developed ape fœtus. The organism, even in the

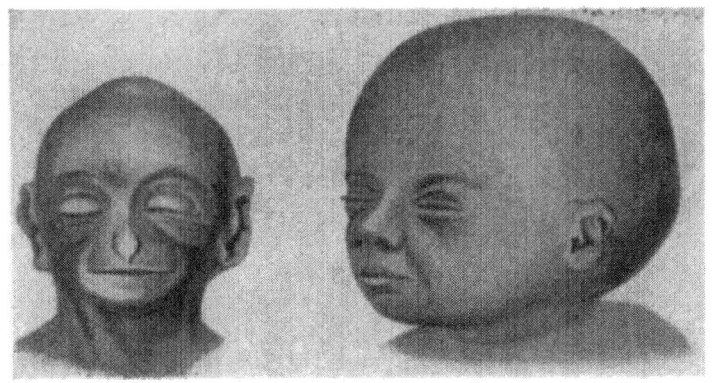

Figure 13.
(*a*) Head of a gibbon embryo (after Selenka).
(*b*) Head of a human embryo of the same age (after A. Brass).

mother's womb, is too matured, no longer youthfully plastic enough, to receive into itself the imprint of an individual " Ego." When the fœtal heads of gibbon and man are placed side by side (Figure 13) this can be seen to an almost terrifying degree. While the human embryo still looks quite tender, plastically soft and round, the ape bears, even before its birth, the wrinkled features of an adult human face. A sinister sight! The animal has aged in the uterus; already it has become almost like an old man. It strove to be equipped too soon, and now the path to becoming human is closed to it.

With Rudolf Steiner's guidance we read in the open script of Nature. Side by side with triumph and progress the record tells of setback and defeat, of failure and surrender. Awestruck, we see revealed in the drama of creation the fateful connection between ascent and fall.

The new doctrine of evolution speaks to Man. If he is not ashamed of his humanity, it speaks even more earnestly and more urgently than its predecessor.

PART III

SOUL

> Animals are fixed ideas incarnate.
> —HENRIK STEFFENS, 1822.

WHAT WE CALL soul is, to begin with, simply a mysterious inner region where outer happenings become experiences. The outer world becomes the inner world. Every man experiences the distinction between the two so clearly that a definition is needless. He knows that everything that takes place in this inner world is his own personal possession. The feeling of colour which is awakened in him when he looks at a flower, the satisfaction that arises in the world of his feelings because of it, and the desire that it can engender (for instance, to possess the flower) are his own affair. No other man can apprehend these directly; another can only discern them in their outer manifestations.

INCARNATION

This inner state, this region of things experienced—perceptions, feelings, efforts—is not present at the beginning of the man's life. It has first to develop. It arises wherever the spiritual world seeks to combine with an earthly body. Of itself, this world has nothing to do with the separation into inner and outer. It is the earth which compels such a separation. " Soul " comes into existence when a spiritual entity " descends " into embodiment.

Supersensible consciousness sees in this process a " self-enwrapping " as it were, a progress through the region which lies between the spiritual and the physical world, and which it names the soul world. When a spiritual being advances to incarnation, it surrounds itself with the substance of this middle kingdom and thus becomes a soul being.

The living organism which is to be the vehicle of this soul being has its origin in a penetration of physical substance with living etheric-formative forces; in this organism physical and etheric body interact. Thus (in the soul-spiritual being), one duality comes down from above; another duality arises to meet it; and an earth being comes into existence through the mutual interpenetration of the four.

The physical body, in itself a mere congregation of the substances and forces of the earth sphere, is made into a whole—into a single texture permeated with living juices—only by the action of the etheric body. Even so, this living product would have to remain in a condition of sleep, if organs were not worked into it by the descending soul-spiritual powers. These organs become the tools for its experience, and its deeds. Soul forces can only be active in the earth-sphere, inasmuch as they "awaken" the bodily organism, and only after they have already worked into the latter the organs for perception and activity.

One need not follow the testimony offered by the seer's heightened consciousness any farther than this in order to realise that *the mystery of the soul* (but only as it concerns man and animal) *is a mystery of incarnation.* The soul-spiritual imprints its seal on the organic-physical. We may therefore hope, with this seal, to find the gateway to the mystery of the soul. Even if we ourselves have not attained directly the power to see supersensible realities, yet with the help of this "seal" we are better able to understand the reports of those who have done so.

The bearing of this on the enquiry undertaken in the present volume is plain, for we are finding that even a physiognomical consideration of the bodies of animal and man offers an insight into the psychic inwardness of each.

★ ★ ★

THE NEW BASIS FOR COMPARATIVE PSYCHOLOGY

This physiognomical basis for comparative psychology was laid down by Rudolf Steiner already in 1910 in a public lecture (in the Berlin Architektenhaus).[1] He uttered the following fundamental statements: "We have to recognise how entirely different is the way in which wisdom is manifested in man and animal. The animal has wisdom in its organs, man has not; man must first acquire it through inner effort."

In endeavouring to make quite clear to himself the meaning of this apparently simple, yet mysterious, comparison, the reader will no doubt remember that the ideas put forward in the chapter on "Form and Shape" are very closely connected with it. There the limb-tools of the animals were considered; and it was seen that these are instruments in a much more real sense than man's practical extremities. Thus, the completion and perfection of the animal limb system was contrasted with the germinal or indeterminate stage which is characteristic of the human hand. The same character was indicated for other parts of the organism. On all sides the animal organs were found to be further developed than the corresponding human organs. Thus, of man it must be said, that he would have remained helpless, if he had not the Spirit, able to turn the unperfected limb-organs to his service, to provide them with ingenious tools, and in this way to replace what Nature had denied to him in his organism.

The deductions from this fundamental law of form, which were made in the first chapter, have already prepared the ground for the new knowledge we desire to reach. We need only recognise clearly how this difference in the bodily equipment—complete in the animal, and incomplete in man—in itself affords a line of approach to the inner life of the two kingdoms. The morphological law of retardation can be extended into the psychological domain.

Let us consider a few outstanding examples. What does it signify, for instance, that the various adaptations necessary for the perpetuation of the race of bees are shared out among particular classes of individuals, and indeed so sub-divided as to be *physically visible*, so that even the unpractised eye soon learns to read in the diversity of their bodily structure a diversity also of

[1] "Human Soul and Animal Soul." Article reprinted (1) in *Die Drei* monthly magazine, Vol. 9 (1929), and (2) in *Gäa Sophia Jahrbuch*, Dornach, Vol. 5 (1930).

function: from the lengthened body of the queen-bee to recognise her role in propagation; where legs are equipped with bristles and little grooves, to recognise the task of the worker-bee; from the great faceted eyes, which are connected together on the forehead, to recognise the drone's part in life. All this gives the impression that the essential part of the creature's activity has been *prepared for in advance* in the equipment of the body. We feel that the drone, for example, certainly has no need to develop attentiveness (a soul quality) in flight, because its head is already so formed that simultaneous perception on all sides occurs as it were of its own accord. The creature does not need to "look around"; those faceted eyes which cover nearly the whole of its head relieve it of the task. If we do not fight shy of the expression, it must be said: the act of looking round is actually inherent in the body. The quality of attentiveness is, as it were, poured out into the very form. Looking further afield, we find that in the ant kingdom the post of guardian is taken over by a special (soldier) class, and these ants are differentiated at once from the ordinary worker ants by a giant head with powerful jaws; these singular creatures can close the entrance to the nest by using their heads as living doors; in addition to this, the work they do *outside* constitutes them a sort of police service for protection. Here again we can see at once that the organism itself has already taken over a large share of the task, in that the quality of defence, flowing out into the appointments of the body, shapes the head into a weapon, the mere possession of which is enough to ensure superiority over the foe.

To borrow an example from the higher animals, one may observe how certain frogs practise brood rearing, and how the adults burden themselves with the spawn. The male toad *accoucheur* wraps the spawn round his legs and carries it about until it hatches out; but the tropical honeycomb toad "Pipa" takes it on his back and the skin seems to be raised up forthwith into small basins, looking like honeycomb cells, which each contain an egg, so that each young frog has a separate aquarium in which to develop. We see in such cases how a function is fulfilled not by any effort of soul, but by a bodily appliance furnished for the purpose. It is the organism itself which makes provision, and, in this singular fashion, relieves the creature from the exertion that would otherwise be needed.

Activities, which have to be accomplished by an effort in the soul of man; which require an expenditure of thought, and which can only be developed through inner striving, discretion, courage, care for posterity, all these in the animal are submerged in the structure of the organs. Sinking a stage deeper, beyond the inner region of the soul forces, they have imprinted themselves into the appurtenances of the body, and become visible form. The very *body* of the animal has become wise and provident.

THE WISDOM OF THE ANIMAL BODY

The essential difference is not that man has more wisdom than the animal, but that wisdom among the animals comes to light at another level of the whole being. Wisdom dwells deeper—within the organism itself. She has taken her place in the animal's bodily outfit.

Thus the animal *finds* the wisdom already there at its disposal, poured out into the organs and irradiating it from them. " We see," said Rudolf Steiner,[1] " in the animal, wherever we may be observing it, that the spiritual reality comes welling out of its organisation. We, who live at a different stage, have to bring in this spiritual reality from all the laws of space and time existence." Spirituality overflows out of the animal organisation; the animal being " enjoys " it. It is actually there, in the animal's bodily equipment, that we can apprehend how that principle which, with such great wisdom, moulds the various organs of the body has penetrated deeper into the animal's organism than into man's.

THE IMPRINT OF THE ASTRAL BODY

For the reader of this book there can no longer be any question what principle this is. It is the astral body, which imprints its seal upon the animal. The animal shape is a physiognomical expression of its astral body. *In it the astral organism is completely*

[1] In the public lectures (previously mentioned) given in the Architektenhaus in Berlin, 10th Nov., 1910.

incarnated, and that principle—supersensible by its very nature—which is the bearer of experiences and impulses has descended into the world of visible things.

But this proposition implies another, which can also be confirmed, when the manifold range of animal form is considered. The animal soul, i.e. its astral body, cannot possess the universality of the human soul. This is expressed in that *onesidedness* of the animal's bodily structure to which reference has already been made. Yet this onesidedness itself has in it something that is truly wonderful. Without reserve the entire wisdom-filled formation of the body and its organs is placed at the service of the impulses or the desires of the soul within. The animal body, with its imposing perfection, is in this respect as if turned out of one mould.

Dependent upon this onesidedness is the fact that the animal forms have grown, so to speak, into their various elements, and have thus grown at the same time to represent the essential quality of the particular element concerned. The quadruped has given itself to be at home on the earth, fish and whale in the water, bird and bat in the air, each suited to its element. If, in addition, the investigator acquires the knowledge which Rudolf Steiner has given of the next higher elements, viz., warmth, light, sound and life-ether (see p. 51) then the task of a future spiritual-scientific zoology comes into sight.[1]

The present volume, however, merely has the task of tracing the essential difference between man and animal. What significance has this deep penetration of the astral body for the inner being of the animal? The attempt must be made to get down to the bed-rock of this mystery, by a kind of physiognomical observation. But here we must not merely consider the shape of the body, but rather try to see it in its connection with the behaviour, " the ways " of an animal. Observation of this kind is unusual, but not difficult to accomplish.

It can be said: the animal possesses wisdom because it possesses its organs. This wisdom, however, reaches out further, and beyond the mere formation of its organs. It reveals itself also in

[1] See the Author's *Tier-Wesenskunde*, second edition, Dornach, 1954.

the way these organs are used. It is there, above all else, in the animal's ways, in its life-habits. To observe these ways, we must be wide awake and on the watch not to confuse them with, nor mistake them for, human practice. Then a reciprocal law is revealed, which was discovered by Carus,[1] a contemporary of Goethe's. *The animal in its behaviour, only continues what is already indicated by the structure of its organism.* These "tools" have been fashioned for a definite use; and this use they reveal already very clearly in their form; so that the uses to which they are put, and everything which results from these, appear to be merely the outcome of the organisation itself. This is in complete contrast to the limbs or organs available for human enterprise, such as, for instance, the hand, which is certainly not prescribed for any particular deed.

Carus (1866, p. 78) speaks of the animal's activity as "arising necessarily out of its structure and organisation.... Its manifestation may, no doubt, have a secondary cause in the shape of certain intensified environmental conditions, but in itself it forms an integral part of the animal life, as in the case of feeding and propagation. The distinction emphasised here is very important; because, inasmuch as we consider the creature's activity as bound up with the bodily structure, it follows that this has nothing to do with any individual power and skill.... For the rest, it is indeed extraordinary that so many products of the wonderful urge to construction may be considered as an actual extension of the animal organisation itself—or indeed *must* be so considered. In this connection we may remember the webs so artfully constructed by the spider—are not these actually an extension of the creature's predatory organs?"

Let us consider (to keep to the example chosen by Carus) what it is that integrally belongs to the spider—the spinning organ itself, the activity of spinning, and the material which is spun. We realise that it is no truer to call the possession of spinning glands able to secrete the necessary matter, and of legs provided with spinner hooks—a part of the organism, than it is to call by this name the skill with which the creature winds together and

[1] K. G. Carus 1789-1869.

draws out the threads, for that " triumphant artifice " the web. The literal *in-fluence* of organic spinning secretion throughout the process is certainly characteristic. Does not this inpouring of the bodily life-sap into the work indicate that the whole performance must be classed as the outcome of the bodily structure? The wasp again gnaws the wood to bits to form a pulp, and then with its saliva works it soft until it can build its paper nest. The bee removes the wax to her comb in the tiny scales which she has sweated from between the rings of her body.

Add to this that all these wonderful powers come into existence in a day, and we get the impression that everything that the animal is capable of is, properly speaking, no less a " growth " than its bodily organs are. The animal contributes as little towards the former as it contributes to the growth of its limbs. Organ and function come to it alike. It consciously acquires neither. It receives them.

Always when these concrete examples are taken, we arrive at the conviction that, *for animals, the birth of particular faculties must be regarded in the same light as the growth of a bodily organ in the embryo.* The chicken, for example, while still in the egg, and as the time of hatching approaches, grows the so-called egg-beak; this it needs to break open the first little window in the egg shell; but when the time has arrived, it knows immediately *how to use it*; the capacity for this has grown with it, as the organ itself has. After the hatching, this growth shrivels and drops off, and the capacity also is lost. Organ and use of organ arise and decay together.

THE APPEARANCE OF CAPACITY

This appearance of a capacity, when the time is ripe, is the primal phenomenon of animal soul development. Wasmann has proved that among the ants, without any sort of learning, the most complicated performances issue immediately from organic necessity; he nurtured and brought up grubs in complete isolation, later on uniting them so as to form a self-dependent colony. And in doing so he observed how the whole range of capabilities was developed among them, without example or training, without uncertainty or error. The sudden appearance of capacity is clearly seen also in the nesting of birds. It is known that many species set about their work with great skill; the small birds arrange twigs and straws into a basket. The swallow moulds basins of clay, the weaver birds intertwine hanging vessels made out of the haulms of long grasses. Moreover this capacity appears immediately the right season has come and operates without further ado, and often far in excess of the actual need. Impulse and capacity for nest building are visibly just as much a bodily necessity as is that very propagation which the nest serves. How utterly unlike human learning! That this quality of capacity belongs of necessity to the body, is shown by the fact that young birds, when building their nests for the first time, are more thorough than the older ones. It is just the opposite with the human learner. With the birds practice does not " make the master "[1] as Wasmann has wittily remarked, but the " prentice hand."

What then really are those capacities that emerge in animals? An invisible growth out beyond the limits of their organs. The final act in the construction of the body passes over into capacity. It is like the blossom bursting out on the plant, when the period of " vegetation " is fulfilled. *For the animals' use of their tools (limbs, etc.) is the blossoming of their organic structure.* The animal as little needs to learn it as the plant needs to learn to flower (Note 19).

If in this way the shape and development of organs and their use are taken together as a single whole, we have before us the working of the astral body as the *one* essential reality at the base of complicated and manifold functions. And then we can understand why Rudolf Steiner describes the astral body as a single member of the living being. And we can understand, too, that originally, and before it united itself with the organism and shaped it in its image, this astral body was in union with the wisdom working throughout the entire Cosmos (thus the name signifies the connection with the world of stars).[2]

---

[1] " Practice makes the master " is a well-known German proverb.
[2] See *An Outline of Occult Science* (1922), p. 410; also *Blut ist ein ganz besonderer Saft*. English edition *Occult Significance of Blood*, pp. 20-31.

LIMITATION OF CAPACITY

This sheds light upon another enigma. The animal organism shaped by the astral body for a definite purpose not only signifies a complete penetration with the forces of wisdom, but also a limitation which can never be removed. The animal must persevere on the way prescribed for it by the formation of its equipment, inexorably to the end. Even its capacities are hard and fast. They are presupposed beforehand by its structure. There is no room for doubt or hesitation, but for the same reason its development is—irrevocably—at an end.

Thus, that which is "instinctive"—i.e. the capacity inborn through the astral body—is maintained throughout by wisdom, but also hemmed in by insurmountable boundaries. The "hardening" of the body which is the evidence of the termination of animal development, is at the same time, when looked at from the soul aspect, a limitation of itself to a particular scene of action, and to those experiences possible therein.

It was formerly said that "the animals are taught by their organs." Goethe accepts this saying, but with characteristic insight, gives it this variation, "the animals are tyrannised over by their organs." Both sentences declare the riddle of "instinct," at once full of wisdom and of narrow limitation. Only a knowledge of the animal's astral body can resolve this strange paradox.[1]

MAN IS BORN HELPLESS

Thus, with the help of Rudolf Steiner's account, we can now see the way to a true description of the gap which separates man's soul from the animal soul. We may start again from intellectual contemplation in the Goethean sense.

Man is born helpless. Other mammals are further advanced than he before they leave the womb. When he comes into the world, his head, his hands, the whole of his upper organisation are still, compared with those of the animals, in a bud-like condition. The proportions of the embryo continue to dominate his body.

[1] The ever increasing literature about animal behaviour contains numberless examples to support this fundamental insight. Cf. K. Lorenz, N. Tinbergen, A. Portmann among others.

From the animal point of view he is a premature birth. And truly, he would soon be lost if he were not helped by the world around him. Nature has left him in the lurch. He is far more helpless than the nestling, which cheeps loudly, and, with eyes wide open, stretches out its beak towards its mother, while the baby at the breast cannot even raise his head.

Nothing that the growing human being acquires, beyond his slender bonus of inborn capacities, comes to him. Everything has to be won. All those abilities, which are the hall-mark of his existence as man, he must make his own through effort. Not for him, as for the animals, can skill and ability be provided with the organs themselves; *he* must impress the necessary ability into them. The human being works into his limbs and organs that in which they are lacking. The ability is not yet " embodied "; he himself must imprint it into the body.

LEARNING TO WALK, SPEAK AND THINK

Rudolf Steiner shows that there are three capacities which the evolving man must incorporate in his body; the erect human posture for walking, human speech, and human thinking. The child is saved no effort. From the first laborious raising of the little head, the first attempts to sit up, to stand, to walk—what a chain of defeats and of undaunted returns to the fray, until at last —" *es geht!* "[1] At last it has achieved what almost every animal at its hatching out or after its birth, almost before it is dry, can do at once. Balance, says Steiner " is already given to the animal's body ", but man has first to learn it. Yet it is for that very reason that he attains the erect walk, which belongs to no other being on earth. It is the human soul's very own activity which must form the organs for its own use.

The same holds good of the achievement of human speech. The cries of the suckling rise from the depths of its organism. Even the first " lal la " is an attempt, its cooing and crowing a learning. From the earliest age of childhood the little being has to strive; the patient listening and beginning over and over again; all these are literally working down—though only in a dreamy half-

[1] The German words mean both " it (the child) walks " and " ça va! "

conscious way—into the organs. It is not the speech organs which teach man to speak; he himself trains them to it. The organs are *developed* through the effort of speech, as every grown man can observe in speech training.

This brings home to us what Goethe, shortly before his death, wrote to Wilhelm von Humboldt (17th March, 1832), "They used to say: the animals are trained by their organs; I would add, so are men; but they are also free to train their organs."

This training is also undertaken in learning to think, which follows upon learning to walk and to speak, and is indeed in clearly defined connection with the latter. Again it is not the actual organ, the brain, that sets the thinking to work—as superficial knowledge might suppose. On the contrary the brain is trained aright only through striving to think (Note 20). Everyone who is not blinded by prejudice can observe that it is he himself who makes the organ a serviceable thinking tool; that he can make it set and hard or living and flexible. In truth man instructs his organs.

Here lies the significance of the early imperfection of the child's body; it must remain open to new influences and to particular impressions, which then can be organised right into it; these are to place it higher than the animal; it is to be a seed which will not ripen as a body left to itself, but through that which must first enter this body. It is a body prematurely born, but with this very incompleteness the way is open to become " man."

The work which is accomplished in the human organism lies open to ordinary sense-observation. Parents and teachers see before their eyes the reality of the Ego at work. They do not need to theorise about it, like philosophers. They observe its work. They watch it struggle with hindrances and overcome them. From within it stamps itself on the outward and visible surface.

THE ENTERING OF THE EGO

This is the phenomenal side of what supersensible consciousness perceives as the descent of an individual soul and spirit-being to take up its abode in a body. Working together with Beings who stand far above man, the human Ego, advancing to the final stage of incarnation, forms the body it needs. It is the Ego which " trains " the organs; the Ego which carries further the life of the soul by giving it a centrepoint to which everything else relates. "The animal is soul, man is spirit. The Spirit Being which in the animal is engaged in soul development has now descended a stage deeper still. In the animal it is soul-forming. In man it has entered the world of sensible matter itself. The spirit is present within the human sensible body."[1]

INDIVIDUALITY AND INHERITED QUALITIES

The human body is nowhere inserted, after the manner of an animal species, into its particular environment. Considered physically, this is a drawback. Considered in relation to the Ego of man, the drawback is actually a gain. It guards him against connection with the world on any circumscribed stage of action. Precisely to this harsh step-mother treatment of his organism man owes his soul-spiritual universality. His inner life, unlike the animal's, does not exhaust itself in the " enjoyment " of the wisdom invested in his organs; it is as if part of this form-creating wisdom, before flowing right down into the organs, had been withheld and placed instead at the service of the human Ego. Because this " I " is not in bondage to a body already fully adapted to the world, it has powers that are free, and able to open directly to the experience of the spirit; with these powers it enters into immediate connection with the spirit, just as through the senses it opens itself to the physical world!

Because of this, man has to receive his body from his parents in a still germinal state, as a " model " (to use Rudolf Steiner's expression) into which he must first build his individual form, upon which he must imprint his Ego (see Chapter II, p. 81). The Ego of man first prepares, by its own work, an abode in the body; it makes the body—as to countenance, movements, carriage and

[1] R. Steiner: "Theosophie, Einführung in übersinnliche Welterkenntnis und Menschenbestimmung", 1904. (English translation: *Theosophy, An Introduction to the Supersensible Knowledge of the World and the Destination of Man*. Revised ed. London, 1954, p. 147.)

outline, nay more, right down into the metabolic system—its own revelation. In the animal body, which merely carries out to its close the metamorphosis already determined by heredity, no Ego could dwell.

ANIMAL GROUP-SOUL

The close connection of the soul-being of the animal with its bodily organisation opens the door to the understanding of yet another mystery (investigated and described by Rudolf Steiner) of the animal soul—its group, or collective, character. The animal soul-organisation owes its origin, not to the individual being but to all those progenitors who have passed it on. Its organs are invisibly, and very closely, bound up with those of its race through the formative-forces body, which body, as the bearer of inherited qualities, links one generation with another; and just as the actual *ability* of the individual creature to do anything arises out of the organism itself, so are all its characteristic qualities drawn from the stream of heredity. Thus, when it is said that the animal " enjoys " the spirit of the world within its organism, it must never be forgotten *that this organism has not an individual, but a racial or group existence.* Thus the influx, as from without, of certain qualities and capacities and their emergence at a certain point in time is an effect of that stream which, flowing downward from sire and grandsire, unites all the individuals of one kind. The animal " enjoys " the use of its organs, " and in the same way, between birth and death, ' enjoys ' its kind " (Rudolf Steiner).

The content of the soul life of an animal cannot be anything limited to an individual; it must be something shared by all others of its kind. Beyond this Rudolf Steiner has affirmed from his spiritual investigation that not only are the experiences of the animal soul herdlike, but that the subject actually experiencing itself—as soul-being common to all—extends over all individuals of the group. He names this the Group-Soul or the Group-Ego of animals. According to his description, in man in each individual only one soul-being is involved; this is the very essence of his being and can be recognised as the Ego. The proper and essential being of the (collective) animal soul, the true centre to which all experiences flow and from which all impulses to action issue, can be designated—to bring it into clear contrast with the human I—as the animal " We." Rudolf Steiner describes the following process. If supersensible sight, after looking at the separate bodies of a genus of animals, advances to the contemplation of the animal's essential being, it reaches an Ego that does not descend into the physical world, but from the supersensible kingdom, holds together the separate animal bodies as though by invisible threads. On another occasion he likened the separate animals to the fingers reaching into visible things from a hand which itself remains invisible. The animal Ego (or " We ") remains constantly in that domain which borders on the sense-world: it is the same in which the Ego of man finds itself in sleep and dreams. It does not enter, as man's Ego does on awaking, into the physical world.

From this description, and from numerous amplifications[1] of it, it can be understood that the animal Group-Egos are beings with a wide-flung consciousness in the higher worlds, but such as only take an indirect share in what happens on the stage of earth. The following description of the form of experience open to these group-souls is applicable only in respect of that which flows to them from the physical world—and from the individual creatures of their group, now made parts of that world. Here, as with all non-human creatures, we have always to distinguish between their own proper significance and that of the beings they make use of as organs with which to experience other worlds.

In relation to earth experiences, the animal Ego owns to the same dependence upon both the inner organic influences and the impressions from the outer surrounding world as man does in his ordinary dream life. Connections can accordingly be perceived between man's condition, when under the influence of dreams, and the *permanent* relation of the group-souls of the animals to the physical world. And everyone can apply them to his own dream experiences.

[1] See, in particular, the Helsingfors Lecture-cycle, 1912, " Die geistigen Wesenheiten in den Himmelskörpern und Naturreichen." (English translation: *The Spiritual Beings in the Heavenly Bodies and the Kingdoms of Nature*, London, 1951. 4th and 9th lectures.)

ANALOGY WITH THE DREAMER

Both the surrounding world and his own organism enter the dreamer's consciousness, not clearly defined as his preceptions ordinarily are, but in the shape of pictures more or less transformed. The dreamer, however, cannot distinguish certainly between the two, for his organic sensations appear no less of an outer world than the other. Organic processes, and particularly organic disturbances, may thrust themselves in very impressive pictures into the dream life and it is only when he awakes that the dreamer can see the connection—for instance, that it was blood congestion which presented itself to him as the room on fire; headache, as a sojourn in a dungeon, full of cobwebs and haunted by hideous beasts. Shut out from the usual sense impressions, the dreamer lives in his pictures. These are not constant; they are continually changing their outline, their expression and their relation to one another. What is most significant of them is the unheeded lapse of one situation into another, the transformation of scenery, remarked partially or completely but in any case probably too late, the appearance and disappearance of the human phantoms. Into this confusion the dreamer feels himself caught up; literally pulled—for quite often he is aware of a dull, even despairing opposition. Everyone knows this feeling of being forced to take part, interpenetrated as it is by a lively sense of powerless willing.

Once awake, however, the sleeper realises that in the place he has come from there was no possibility of taking the situation in hand—nothing but an obscure impulse of longing and antipathy begotten by magical spellbinding pictures. How often it astonishes him that in the dream he was completely at the mercy of some situation whose absurdity is now transparent. He found himself in an entirely abnormal relation to his surroundings—and took it as a matter of course. For instance, he could not go on walking—must have slid helpless to the ground, had it not occurred to one of the many passers-by (yes, but *who* was it?) to help him. But in the dream itself he did *not* get annoyed and think how nothing of the kind had ever happened to him before, no; he concentrated all his efforts on making his pitiable plight known, uttered choking cries for help and so on. It is the fact that in dreams action arises, as it were, out of the situation itself, and without the dreamer forming any judgment on it. He is delivered over to his experiences, and his willing is but a blind unreasoning drift.

FIXATION

The dull fixation of consciousness by means of which, in the dream, an imperfectly grasped experience brings about a state of affairs, is exactly what we observe in animals. With them too, there is awareness of circumstances rather than penetration of them, and this awareness leads direct to action.

If the animal Ego does not really step down into the world of waking life; if it is really " absent," then the fixation of animal behaviour becomes entirely comprehensible. And we can also comprehend with what intensity of fear or of desire it meets these situations that it does not understand. It never gets near the region where awakened judgment and conscious decision come into play.

According to Rudolf Steiner's description,[1] animals which have no voice in the sense applicable to the warm-blooded animals live *entirely* within this picture world—(the higher animals will be considered in the next chapter). In place of an " outer world " these lower animals only know the intensely and symbolically experienced picture. Their experience is like that which was possessed by the human race in the " Old Moon " (see Chapter II, pp. 50, 51). Rudolf Steiner puts it as follows: " It is true that the resemblance of the images in the moon-consciousness to those objects to which they are related is even slighter than that of our dream pictures; but to make up for this, there is a perfect *correspondence* between picture and object. At our present stage of earth evolution the point to notice is that the representation or idea is a copy of the corresponding object. This is not so in the moon-consciousness. Supposing, for example, that the moon man draws near to an object which is sympathetic or advantageous to him, a colour-image rises up within his soul of a clear bright

[1] Cf. following p. 109.

quality; if something hurtful approaches him, he has an ugly gloomy picture. The representation he sees is not a likeness, but a symbol of the object, and this symbol corresponds with the object in a quite definite way, and according to fixed laws. It follows, that a being who experiences sense-images of this kind, can direct his life accordingly. Thus, the soul life of our moon-ancestors took its course in pictures—pictures which resembled those of present-day dreams in that they were of a fleeting, hovering, sensuous nature, but differed from these in their perfect subjection to fixed laws " (*Lucifer Gnosis*, 1905, No. 29).

Let us take some well-known examples of dream symbolism in order with their help to transpose ourselves into the world of the animal. A whirring alarm clock may be experienced as a train roaring past, and the falling of an object as a cannon shot. If the feet become heated, the dreamer sees himself stepping barefoot over warm lava-streams, if a straw gets in between his toes, it turns into an arrow piercing the foot and holding the sufferer fast, and so on. The awakened victim discovers his mistake, but the very idea that he could make a mistake was unable to reach the dreamer!

" MISTAKE " OR DREAM?

The same incapacity to understand substitution is evident in the animal directly it is removed from its accustomed conditions to strange ones. Then it squanders the greatest pains on entirely useless actions. Thus the wasp, when it tries to fly through the window-pane, or the ant-lion when it makes despairing attempts to work itself backwards into the table. Especially convincing is an observation, which is to be found in André (1924, p. 42): if a spider's net is touched with a vibrating tuning-fork, the spider hurries up and spins a thick covering round the metal point, as if it were a fly to be stopped from escaping. It is obviously foolish to treat this as if it were a human occurrence and say the spider " confuses " the tuning-fork with a fly, for it has certainly never seen the tuning-fork at all as an " object," nor does it know any such thing as " fly "; instead, whenever the net trembles, a picture is called into existence, which lays hold of the spider's being with the same magical power, and impels it to action, in the same way as the sensation of the stalk pricking the dreamer's foot impelled him to flight. This example may be allowed to teach us, how exactly Steiner's description reproduces the inner world of such animals, and how little similar this world is to the world which surrounds awakened man. We really must cease once for all to imagine that the inner world of the animal is poorer than that of man; it has a gay variety, and liveliness, beside which our human world of mere "things" looks simply prosaic and beggarly.

Animal behaviour, with its peculiarities, is perfectly comprehensible when Rudolf Steiner's account of the way in which the animal Ego does not abide on the physical plane but in the neighbouring astral world is once grasped. The accompanying scheme to which he often referred (see diagram), can, if used aright, form the basis of an animal psychology.

	Mineral	Plant	Animal	Man
Higher spiritual world	Ego			
Lower spiritual world	Astral body	Ego		
Astral world	Etheric body	Astral body	Ego	
Physical world	Physical body	Etheric body Physical body	Astral body Etheric body Physical body	" I " (Ego) Astral body Etheric body Physical body

Diagram 14.—Stages of Incarnation of the Four Kingdoms of Nature (after Rudolf Steiner).

ANALOGY FROM THE ANIMAL FORM

In conclusion we may perhaps endeavour to make clear by an analogy how the incarnation of an animal takes place.

All experiences which have the clearness of man's daytime consciousness have their centre in the upper pole of his organism (cf. pp. 15, 16). According to Steiner's description the living formative forces are there so far extruded from the organic substance that this upper pole can only continue to live because it supports itself on the rest of the body, and is sustained by it. The head is really continually dying. And, as Rudolf Steiner recognised, only by means of this process of dying is it possible to be awake. The medial organisation, centred in the breathing and circulation system (chest), is a degree more living—but for that very reason it has only a dream-like consciousness. Only the lower region of man, particularly the general system of his limbs, is completely dominated by the life processes: here the mobile plastic powers able to build up, to form, are working to their full extent; it is from here that the upbuilding and the maintenance of the human form is carried on. But from this very fact it follows that here no consciousness is united with these processes: for life and consciousness are opposites (see Note 21).

It is only in those experiences which are sustained by the wakefulness of the head-system that freedom can rule. Side by side with these, or really beneath them, there are human conditions in which neither awakened judgment nor individual decision is working—where, instead of these, there is only a dreamy awareness and a dreamy impulsion. Everyone knows such cases, where decision is made and the thing done out of a consciousness that is only half awake. Everyone, who observes himself, knows how countless transactions of everyday life are carried out in such a way that, though the spring of action is withdrawn from control, yet they are not entirely unconscious. One may sometimes, for example, detect an earlier experience—generally a painful one—in the very act of passing over into a half-conscious " opinion " and we are certainly justified in suspecting that we often unconsciously throw the entire experience of recent years into the balance, when we " weigh-up " (*erwägen*) a situation, as the genius of language so strikingly puts it. On such occasions it may well be a long time before we recall all that we have thrown into the balance.

Upon subsequent—and *waking*—reflection the opinion thus formed often reveals itself as a prejudice, an emotion, or merely a kind of laziness posing as the outcome of thought. Incidentally, so-called judgments on the subject of Anthroposophy can be given as examples of this. There takes place in the region *below* the waking consciousness a secret argument in which the true and the false chaotically run into one another. Brought up to the light and formulated, much of this proves convincing; on the other hand much is revealed as mis-shapen birth of all kinds of desires. (Freud in his *Psychology of the Daily Life* has furnished some brilliant observations; only it is a pity that he has made of the human soul in which he comprehends only the vices, a being of such dishonesty and vulgarity.) Thus we can actually speak of a sagacity of the dream-region; but with regard to this region—which corresponds morphologically with the chest organisation, physiologically with the breathing and circulation processes—it must be remembered that influences from the body play a vigorous part in its experiences. The dreamer's " cognition " of the world, as against " recognition," and the dream-like " behaviour " of animals, in place of human dealing, must be thought of as anchored in this region.

Further down, in the region of the regenerating, upbuilding forces of the lower organisation nothing can be found that would correspond to the wakeful judgment in man: instead these parts of the body are the receivers of formative impulses. Here, as instanced in the preceding pages (p. 94), the astral body, as the mediator of the model or race-type, is actively building-up: and behind this again there dwells, in man, the Ego, with its own impelling force.

Varieties of experience, which are not firmly taken hold of in full consciousness, must always tend to be " submerged." They are forced to seek embodiment in the region of the upbuilding forces, and there to work formatively on the body. All that could not be grasped in self-consciousness, all that falls below the experiences open to the human Ego, the more vigorous it is, the more it must seek expression in actual bodily form.

For the animal this failure to take hold of experiences, which then sink down into the bodily form, is, of course, the normal thing. When this becomes quite clear to us, it opens up an amazing insight. Rudolf Steiner, in the lectures held in 1914,[1] and published in English 1927, "Ways to a New Style in Architecture,"[2] has already cited a significant example. He speaks there of animals that live in the sand, and says of them that with their astral body they enter into a very close relationship with the coloration of the sand. " The consciousness ' colouring of sand ' flows through its whole being. And the being takes on the colour." Such knowledge as this cannot but have extraordinarily far-reaching consequences. The so-called " adaptation to environment " is an essentially animal phenomenon and has its foundation in the astral body's close connection with its environment. In man the connection is quite a different one. " All animals," says Rudolf Steiner, " live as it were under the surface of the sea of colour and light. . . . *Man with his Ego-consciousness stretches out beyond the sea of colour and light,* and the very fact that he can do this gives him his Ego-consciousness." A colour such as red is changed for us into a stream of sensation, just because with our " I " we can emerge from the streaming sea of colour itself.

THE SECRET OF SO-CALLED ADAPTATION TO ENVIRONMENT

How protective colouring is provided and how other so-called "adaptations" occur, are some of the great riddles in biology. Here they are referred to the special character of the animal astral body. Such manifestations as these can only come into existence because the astral body of the animal is not interpenetrated directly with an Ego (as the human astral body is): instead the animal is in the strong grip of its experiences, dreamy, yet vivid, as these are.

Precisely because the astral body of the animal is closely bound up with its organism, its experiences imprint themselves physiognomically into its shape. They sink down into the plastic regions of the body (into the metabolic system), and here, by means of that etheric linking of body to body which is the real " group," they are communicated to the other creatures of the same species (particularly to those of the coming generation). Because of this

[1] " Wege zu einem Baustil."   [2] p. 45.

submergence these experiences are not elaborated and used up as soul, but instead are worked over into the body. Thus it is that the animal body is organised into and adapted to its environment.

Man, when alert and healthy, inwardly digests and refocuses his experiences (cf. Chapter V, p. 143). Thereby he ripens in spirit and soul. In the case of the sick man this activity of the Ego, preventing the submersion of experiences into the physical and raising them to the level of soul, may be suspended. And when the body is no longer protected against those images contained in the astral body which are constantly descending into it, a state of disease may set in such as the following.

THE ANALOGY OF HYSTERIA

Kretschmer describes in his book on Hysteria (1923, pp. 113-114) an appalling case in which it can be seen how experiences that are not directed by the Ego, may bore their way, by degrees, into the body. " An unhappily married woman, for many months had one desire. To get away from this house, to get away from this marriage." This desire forms the ground-tone, the " motif," of her personal attitude throughout the unhappy union. Next this leading motive begins to assume " pictorial imaginative forms " (i.e. it sinks into the dream-region and is no longer clearly grasped). " Escape at night, running away and jumping over hill, dale and forest. Then the psycho-motility begins to work . . . though in a semi-conscious state " (i.e. the experience has by now descended into the region of the limb system). " Attacks of Poriomania break out, with blind efforts to run ' away.' At last (and now it is experience no longer, but a form-giving impulse which has burst through into the body's upbuilding processes) symptoms of rigidity in the limb system are developed; the gestures of flight and running away have petrified into an expressive gait and carriage of the body, which represents the ruling motive ('If only I could get away') in a sort of permanent ' mime.' "

Thus by considering a human being in a state of *soul sickness* we obtain an insight into the psychic condition of an animal. Just as this unhappy wife, paralysed by her past experience, and unable to use it for her own development, allows it to sink down into her bodily formation—so the animal group-soul, confined to the experience of its accustomed environment, must because of its close connection with the body busy itself incessantly about imprinting the ruling motive more deeply into the physiognomy of the animal's body. The animal soul never gets beyond this connection with the past, it never dominates the situation, is never " on the spot," never where " it counts " to be. It is as if it were dreaming in its world about the day before yesterday—always as if it had arrived too late. *It has not spiritual presence of mind.* Animal behaviour is haunted by the shades of its ancestors. It does what it has been decided, before its birth, unconsciously, by its species, that it shall do—and this is the true " pre-judice." Its impulses are as though half forgotten, or not quite within the reach of memory; it is as if, having already decided something, it were unable to remember what and why, as if it were acting on the strength of a lost tradition, whose meaning it no longer understands.

This is the heart of the difference between it and the human soul. The latter may indeed degenerate into such conditions. But its true being lies in dominating them.

PART IV

EXPERIENCE

What the animal lacks is a single centre, enabling it to relate its functions of seeing, hearing and smelling to one and the same concrete thing, to one identical core of reality.
—MAX SCHELER.

THE ATTEMPT TO distinguish the animal's world of experience from man's obliges us to transcend the traditional standards of scientific enquiry. Thus the reader is here referred even more than was the case in the preceding section to his own powers of perception. Conviction will depend not so much on exterior proof as on the intrinsic self-sufficiency of the whole presentation. That is not to say that the system, once it is grasped, does not require to be confirmed by experience. Although it can only carry in itself the proof of its authenticity, it will nevertheless be possible to demonstrate at any point how everything that is said is borne out by actual observation of men and animals.

ANIMALS WITH AND WITHOUT TONE

Rudolf Steiner distinguished two animal kingdoms. In a course of lectures given as early as 1907[1] he remarked that spiritual investigation acknowledges a deep cleft between those animals which cannot, like human beings, give utterance to *tone* from within themselves and the others which are endowed with this tone faculty. The first, or toneless, animals still possess in the present day that pictorial consciousness proper to the Old Moon of which we spoke in the previous chapter (p. 101-2). They do certainly produce tones and noises but only with the outer parts of the body—for example, the chirping of the cricket, cicada

[1] " Die Theosophie des Rosenkreuzers"; English translation *The Theosophy of the Rosicrucians*, lecture 8, London, 1953.

and grasshopper. Nor do even the cold-blooded vertebrates, e.g. frogs, emit sound in the human way; their voice is a kind of rattle in the cavities of the mouth or of the throat, as the case may be. It is only the warm-blooded genera (birds, mammals, man) which achieve genuine intonation; for the reptiles, as the hiss of the snake proves to us, do not really rise to it.

All the lower animals then, and, among vertebrates, the cold-blooded, have remained behind at that stage of symbolical picture-consciousness which was compared above (p. 100) with the dream world; save that in it (according to the description given by Rudolf Steiner, to which we have already referred) an unequivocal correspondence between object and image took the place of the kaleidoscopic symbolism which marks our dreams to-day.

Let us consider once more the case of the spider. She sits waiting in her web. Touch it with a vibrating tuning-fork, and she will spin round the tip of that instrument as though it were a captive fly. Obviously there has been no perception here of a "thing" (fly) in the sense in which human beings recognise separate objects in their environment; instead, we have to do with a *summary awareness of the situation as a whole*, an awareness which, apart from human interference, would have been perfectly adequate to induce correct behaviour. The dominant factor in the creature's field of experience is no doubt the violent quivering of the web, but many other contributory details must go to make up the whole experience. There does not, however, stand out from this whole, as would have been the case with a human being, one *thing*, a fly; and there cannot therefore be said to have been "comprehension." Nothing but such a defective experience of identity will explain the surprising helplessness displayed by creatures of this sort in situations which to the human intelligence appear to differ hardly at all from the normal. Hans Volkelt, for instance, observed that if a fly is proffered to a spider *outside* her web, she will start back with every indication of fear (violent movements of revulsion, etc.), just as if she did not recognise the very prey which, had she encountered it *inside* the net, she would so speedily have overcome. Here too there is, unquestionably, no projection of a single object (victim) from the creature's environment. What happens is the arising of an interior picture—in this case a terrifying one—by which its behaviour is directed precisely in the manner depicted by Steiner as true for moon humanity (Note 22).

TONELESS ANIMALS IMPRISONED IN THEIR PICTURE-EXPERIENCES

Summary perception of the situation as a whole—we are driven to it again and again as the one possible explanation of the singular behaviour of the lower animals. There is no other way of understanding the much discussed " errors " of instinct, as they are called. Dragonflies, for instance, usually lay their eggs while flying over water. When they deposit them on a fresh tarred barracks roof instead, the mistake is gross enough from the human point of view. It becomes intelligible if we assume within the consciousness of the insect, not a sort of copy of an articulated and shapely environment but one complete *single* experience of a sensory nature, an experience of such a kind that in it the separate components (separate for *human* consciousness) are gathered up into a unity. And summary perception of this kind may be conceived of as by its nature symbolical; for a symbol, in so far as it is the real thing and not a mere allegory, is something which condenses all the richness of a manifold content into one representative general impression.

INSTINCTIVE ACTION IS A STRIKING THROUGH OF THE PICTURE-EXPERIENCE

Passing to behaviour, we find that such a conception of the imprisonment of the creature within its picture-experience helps us to understand how perception may issue immediately in physical event. The intensity of the experience as we know from our own dreams (see previous chapter, p. 101) renders such an immediate striking through into the sphere of action credible to us. The impression vents in physical activity virtually without a moment's interval; it streams without obstruction into an exertion of the organs. This accounts for the delusion under which so many modern students of animal psychology labour, that the instinctive process is a mechanical affair somewhat like a stop-watch. They would like to make the automatic reflex the prototype of animal behaviour and the animal itself a complex of interacting events, which could take place equally well without consciousness.

But we have no right to think of the animal's world of experiences as being in any way poorer or less complete than the human being's "world." Even if there is no distinct focusing of "things," still it is quite as various, quite as much shot through with suggestive qualities, as the human dream-world, about which there is most decidedly nothing penurious, though in the degree of consciousness it is inferior to the waking state.

Of what nature, then, is the behaviour of lower animals? Is it rigidly prescribed, even to the details?

PLASTICITY OF INSTINCT

For a long time it has been remarked as an anomaly by experimenters that the working of instinct is at the same time rigid and flexible. The ant-lion will throw up sand with the utmost skill over an ant that has fallen into its artificial hole, thus precipitating its fatal fall to the bottom. Yet the same creature on a dinner plate will exhaust itself in impotent efforts to dig itself in backwards! It appeared to survey the ant's fall quite intelligently, but now lacks all perception of the senselessness of its behaviour. It is significant, too, that precisely those insects whose wonderful community arrangements excite the admiration of man should be the ones in which we find this contradictory behaviour most pronounced. Within the limits of their own physiological horizon these creatures will even *profit* by experience, inasmuch as they will modify a given operation very considerably, to suit new conditions; while it has actually been shown by Wasmann that the behaviour of ants is more constructive and more adaptable than that of birds. And yet the same ant, Polyergus, which will attack an enemy nest in closed columns, killing all that make a stand but allowing the fugitives to escape unharmed once it has secured their *larvæ* and *pupæ*—this same ant will starve to death in the immediate vicinity of ideal provender, not because it cannot feed itself, but simply because it lacks the most rudimentary capacity spontaneously to associate the sensation of hunger with the perception of food; only if its lower mandible accidentally comes into contact with the food will it actually devour any; for in the ordinary way it reckons to be fed by the raided " slaves " (Wasmann, 1900, p. 32 *et seq.*).

Here then we have, humanly speaking, cleverness and resource on the one hand and, on the other, obduracy and limitation, both in one and the same creature.

Up till now conventional animal psychology has been in the difficult position of having to evolve an idea of these creatures' experience such as would make both kinds of behaviour intelligible, the wisdom-filled within the bounds of the usual situation and the apparently stupid outside those bounds. The comfortable hypothesis, based on certain observations, that an animal's " instinct " is of the rigid and limited kind which prescribes one definite action from the performance of which it can by no means escape, is continually being upset by other observations.

The means which an ant, for instance, may adopt in a particular case in order to reach a particular objective, will alter with circumstances to a degree which puzzles the layman altogether. It can take the most different ways, employ the most diverse means, overcome or get round the most different obstacles, in a word—it is very far from " blind," but on the contrary takes advantage of every available circumstance in order to make the best it can of a situation. Moreover, each particular animal may act differently, finding its own " individual " solution. When Wasmann put a salamander into an observation nest of *Formica sanguinea*, the ants had in a short time completely walled it in with earth; on the back of a beetle they laid lumps of earth, until it could no longer move.

So far then from always presenting the same pedantically stereotyped appearance, the behaviour of invertebrates is marked by a striking plasticity. There is no question of one definite process being set in motion, which must lead to a definite result; on the contrary there is very considerable free play. The arrangements which these animals will make depend on the particular situation at the moment; only the final objective appears to be determined in advance. The outward circumstance in no wise produces, like an electric switch, a mechanically answering effect; rather it selects from the wealth of possibilities one particular modification of the animal's ingeniously directed behaviour. There is no objective continuum of necessity from "stimulus" to behaviour; an intermediate region has to be traversed, in which the chain of mechanical causality is broken. Here, at the transition-point between experience and action, that is to say, at the point where the picture-experience issues in an impulse to act, there must come into play a certain faculty of the animal soul, which relates the more general tendency to the particular situation. The content of the impulse is fixed all along, but, since it is plastic, its form remains in suspense, until it has been fashioned into a special variation adapted to the circumstances.

BEHAVIOUR TYPES

Is there not an obvious analogy here to the relation between a morphological type and its environment? When a given plant is transplanted, say, from a plain to a mountain, it varies in a certain definite direction; but the variation itself is no less peculiar to the genus than are the *differentiæ* which determine it; another kind of environment evokes yet other modifications; until gradually, by multiplying the different external conditions, the invisible realm of innate tendency can be step by step explored and marked out. Out of the stock of imperceptible qualities first one and then another is rendered visible, and has henceforth to be assigned to the genus as one of its properties. Goethe's "type" is that entity which is the ground of all these properties, and from which the particular variation is educed by circumstances. The circumstance does not create the type; it only makes it visible; it assists it to manifestation.

In the same way in animal behaviour each particular case is drawn by environmental conditions from a hidden exchequer of plastic impulse. It is potentiality made visible. Animal behaviour does not slavishly copy hard and fast patterns; it is grounded on types capable of variation.

Now we begin to see light on that obscure phenomenon, the

"plasticity of instinct." If every action is grounded in a general impulse, modifiable formally but predetermined as to content, it is surely obvious that this plastic type on the action side must be closely connected with the "summary awareness" on the experience side of an animal's nature. Clearly the two together form the twin poles of one indivisible process. At one end the picture, uniting the details in a single experience, at the other the *type* of animal activity, stamping its seal upon a mutable environment.

THE INWARDNESS OF ANIMALS

The connection between this last and the "dreamlike knowledge and action" of the previous chapter (Note 23) is clear enough. Knowing and doing, which for human beings are torn apart into these two halves (we shall examine it more closely in a moment) coalesce in the case of the animal into one single process, the whole being under the predominance of the one picture-experience. And this experience vents in action in much the same way that the experience of hunger passes over, at the sight of food, into the secretion of saliva. The plasticity of animal behaviour and the restrictedness of animal behaviour both have their roots in this peculiarity of their nature. The wasp—"usually so clever"—provides the classical example for the limitations to which instinct is subject by the way in which it goes on trying to bore through the window pane. A moment ago it hit on some almost refined device to enable it to carry out its usual arrangements; now the palpably stupid creature is at its wit's end because of some absurd little alteration which is demanded of its behaviour. After finding a hundred solutions, getting round a hundred obstacles, it boggles at the last and (to us human beings) quite negligible step. Suddenly a barrier has arisen and the phantasy, whose inventive power seemed to be so sustained, fails of the little more it needed to overcome this difficulty too. *It is the centre of activity, not its perceptible outer fringe, which is rigid and unaccommodating.*

There is one more consideration, however, which is necessary to a proper understanding of animal behaviour. This was recently compared with glandular secretion. For there too it may be observed how a physical process is sympathetically experienced, if only dream-fashion, as a kind of disburdening or relief from pressure. The physical activity is enjoyed in its underlying processes, it is *savoured* by means of the consciousness, which, with a hint of exertion about it, still lingers on in such processes. Thus every deed performed by an animal is enjoyed by it as a kind of explosion, as a relief from the pressure of congested forces, and to that extent it may properly be compared with the sensation of

Picture Experience
(*Overlapping*)

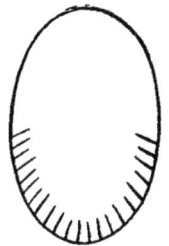

Physical Process
('*Savoured*')

Diagram 15.
Diagram of the inwardness peculiar to " creatures of instinct."
The hatched portion is intended to indicate the dimming of consciousness in the part concerned.)

pleasure which a man feels when a gathering weight of suppressed annoyance at last breaks out in a storm of hearty abuse, and he is able to " breathe more freely." In such a case there is really something of that pleasurable " resolution " which Freud was nevertheless wrong in holding to be the basis of all human actions. What it does give us is an insight into the purely *animal* part of an action, that part which Thomas Aquinas described accurately enough long ago, when he said that " the activities of beasts are pursued for the sake of pleasure " (Operationes quaeruntur propter delectationem), thus pointing even then to the extension of the animal's consciousness into the physical processes that effect outward action.

To sum up—the inwardness which is realised by the lower animals is dominated at the percipient end by a picture experience having no direct relation to individual objects, while at the opposite end it vents in a physical process grasped, or rather "savoured" in a dreamlike way by the animal's consciousness.

The diagram on page 116 is intended to illustrate the polarity. It may also serve as a basis for the comparison, which follows, with the corresponding process in human beings and the higher animals.

CONSCIOUSNESS OF ANIMALS CAPABLE OF INTONATION

In order to gain an insight into the consciousness of the higher animals, that is to say, of animals able to give utterance to tone from within themselves, we may start from a different point altogether—that of human consciousness. For it is possible to understand the former by working downwards from the latter, just as the forms of the mammals can be understood by regarding them as degenerate modifications of the human form. The first chapter of this book demonstrated a progressive morphological descent from the central Human Form. Now the representation of human consciousness as it really is, is to provide the prototype from which animal consciousness may, through a kind of degeneration, be deduced.

A comprehensive account of human consciousness, starting with no assumptions, is to be found in Rudolf Steiner's *Die Philosophie der Freiheit* (1894).[1] To the extent that the description given there of the elements of human consciousness is transferable to the consciousness of animals, that book, strange as it may sound, is really the foundation-stone of a spiritually scientific psychology of the higher animals.

STRUCTURE OF WAKING CONSCIOUSNESS

Rudolf Steiner describes the consciousness of waking human beings in such a way that every unbiased person can verify the truth of it upon himself. For Steiner takes as his starting point the

[1] English translation: *Philosophy of Spiritual Activity. Fundamentals of a Modern World Conception.* London, 1921, and later revised editions.

one thing in the world that is immediately "given"—viz., human thinking. *Every* item of knowledge is fastened, as with a secret umbilical cord, to this original centre. Steiner's exposition therefore takes its start from a first principle which can be verified at any moment to be such.

A chaos of unrelated percepts would be all that is given to us, did there not arise in ourselves an interior activity, by which the chaos is reduced to order. True, this activity is itself "given," like everything else, but it has at the same time the quality of leading out beyond the merely given; for in our thinking something interpenetrates us which has its own laws and which at the same time reveals the laws to which the rest of the world conforms. Out of the "disjecta membra" of mere sensible appearance[1] thinking creates an articulate whole.

"The world at this stage of our consideration must be conceived as a perfectly flat surface. No one part of this surface stands out from the rest, or shows any conceivable difference from another. It is only as thought strikes its spark into the surface that heights and depths are introduced, that one part appears to be standing out more or less above another, that everything takes on definite form, and threads twist about from shape to shape."[2]

ACTIVITY OF THE EGO IN KNOWING

To begin with, then, there is given to human consciousness merely the *sum* of the details afterwards to be perceived; but it has in itself the power to reveal the network of their relations, their connection and disconnection, their hierarchical organisation into ever larger and larger wholes, their emergences from and relapses into one another, their antitheses, their affinities and repugnances, their opposition, competition and mutual aid. It does not do this in any arbitrary way, but simply in its process of pene-

[1] This expression "Erscheinung für die Sinne" was first used by Steiner in his *Grundlinien einer Erkenntnistheorie der Goetheschen Weltanschauung* (Introduction to a Goethean Theory of Knowledge), which appeared in 1886. English translation. *The Theory of Knowledge Implicit in Goethe's World Conception. Fundamental Outlines with special reference to Schiller.* New York, 1940.

[2] Ibid., p. 18.

trating deeper into the nature of things than perception, directed upon their outer surface, can do. Thus, it reaches that level of reality at which man lives in a vital connection with things instead of merely seeing himself set over against them. This is the all-permeating, imperceptible world of formative forces. Man is embedded in it together with all things and, though he does not know it, it is an echo of this unconscious commerce that arises in his consciousness, when he thinks. He can prove to himself that he is so embedded, as soon as he can convince himself that in intellectual apprehension he restores to the world of his perceptions something which he took from it as long as he merely perceived. In the act and moment of knowledge he has reinstated the dissevered half of reality[1] and set up once more the *original* coherence.[2]

The one great difficulty is that the restored element reaches our consciousness in a pale and emaciated form and not in its genuine vital activity; it reaches us as the concept. In this way the illusion arises that man is importing from " within," into a world which is generally speaking given " from without," something that is foreign to it. But the point is that he himself has previously —unawares—divided the world into two halves. Usually percepts are regarded as being objective, as something which would be there just the same without any human co-operation; but in reality the perceived only comes into being through the fact that man abstracts from the world something which he afterwards restores to it in his thinking. To perceive is to take the life from the vital warp and woof of being; to know is to give back to things, by restoring the concept, the vital and organic structure which constitutes them what they really are. In the world itself there is neither percept nor concept, for these only come into being through a dichotomy which man himself, owing to his peculiar organisation, effects; they have no significance except for man, for " it has nothing to do with the nature of things how I happen to be organised to comprehend them."[3] The cleft,

[1] *Philosophy of Spiritual Activity* (English edition 1921), p. 87.
[2] Ibid., p. 91.   [3] Ibid., p. 82.

which man by his organisation has split, is bridged over again by the activity of his Ego.

ORIGIN OF CONSCIOUSNESS OF OBJECTS

This cleavage and the overcoming of it are an essentially human experience. Man must keep on going through them afresh, for they are necessary to his unfolding of an independent consciousness. If he were given over in a dreamlike way to every occurrence in his environment, as the child still actually is, he would be unable to experience himself as " I," contrasting his ipseity with the world. The child participates, as it were, not only in the activities with which he is surrounded spatially, but also in the psychological states of his parents and teachers; he is, especially in his first years, " entirely a sense-organ," as Rudolf Steiner pointed out (1921).[1] He receives in the same measure impressions of sensible and of supersensible events, and imparts to them in his turn the influences of his own still plastic system. When he sees objects, he lives into them without detachment, not yet contrasting them with himself. He knows as yet neither an inner nor an outer distance from things—will grasp at the moon and blow out the rising sun like a candle. He lacks the "bird's-eye view." His world is still in living solution; it is a flux, out of which separate things will only crystallise later. But gradually, as the distance of things from one another is experienced, he comes to feel in like measure his own isolation from the world. Corresponding to the external disjunction of " objects " goes a personal withdrawal from them. Perspective and vanishing-point are born for the first time in his experience and with them the possibility of reproducing it in black and white. This vanishing-point is the counterpart of the Ego's subjective " standpoint " in antithesis to the world (Note 24).

In the same way percept and concept are gradually sundered for the human being as he grows up. The child's perception is still saturated with spirit, or mind; his thought is still a kind of " intuition " (intueri—to gaze at). No need as yet for that

[1] Der Lehrerkurs R. Steiners am Goetheanum, bearbeitet von Albert Steffen 1921. English translation *Lectures to Teachers*, London, 1948, Ch. VI.

" synthesis of percept and concept," of which adult knowledge consists.[1] Not that he would be too stupid, but simply that for him the cleft between the two, which beckons us adults on to knowledge, does not yet exist. The hiatus between thinking and perceiving is not yet there. It " only arises in the moment at which I, the contemplator, stand over against the thing contemplated."[2] Whereas the child neither " stands over against " nor contemplates. " The strangeness of an object lies in its detachment,"[3] and the child knows nothing yet either of objects or of detachment. Everything is still in the original unity, from which it is only afterwards " torn by the nature of our organisation."[4]

THE ACTIVE ASPECT OF " OBJECT-CONSCIOUSNESS "

On the knowledge side, then, the waking consciousness of a human being depends on his becoming aware of reality in a twofold way. On the active side, as Rudolf Steiner laid it down in the *Philosophy of Spiritual Activity*, that which is specifically human is a kind of moral creation—underived, independent of organic occurrences and needs. It is moreover the translation of this intuition into action. In action the subject tears itself free of its connection with the processes of the organism.

Were man to participate in these processes with his full consciousness, it would be beyond his power to compass an autonomous impulse. There would always be the danger that some organic event or natural necessity might worm its way into his decision and threaten the absolute nature of the moral concept. Now in all the operations of desire and instinct this interference is a matter of course and not therefore open to reproach. But where one may fairly speak of "an absolutely original decision,"[5] there physical stress and physical needs play no part, and the very thing which is the whole significance of animal exertion— the " *delectatio*," *or physical delight in action*—is omitted from the reckoning. In human consciousness there can be no more than a picture of the deed born out of the independent creative power of the Ego: the impulse to its realisation is the last thing that

[1] *Philosophy of Spiritual Activity*, p. 86.   [2] Ibid., p. 83.
[3] Ibid., p. 91.   [4] Ibid., p. 91.   [5] Ibid., p. 198.

is experienced with full intensity; all that follows sinks below the level of waking consciousness, and the actual setting in motion of the body takes place in a realm over which darkness is spread.

Thus, on the one hand, the human Ego experiences itself as contrasted with a world divided into two halves, percept and concept, while on the other, it withdraws, as far as its conscious experience is concerned, from the organic processes, leaving these to the deeper strata of the soul. How all this expresses itself physically in the raising of the head and the sinking of the limbs which bear it, has been told in the previous chapter (p. 101). " We are truly men, only in so far as we are free."[1] Head erect, limbs given over to the earthly gravity—these are the foundations prepared by destiny for such a true human Being.

THE EXPERIENCE OF HIGHER ANIMALS ORIGINATED IN THE EXCLUSION OF THE EGO

We have now reached a position from which to understand the experiences common to the higher animals. Eliminate from the idea of human knowedge and behaviour, which has just been developed, everything that comes about through the activity of the Ego, and what remains will apply to the higher animals.

The daytime world of a human being is the world of objects. He calls it, with some justification, " the world " pure and simple. It was shown above how awareness of an environment consisting of discrete things is grounded on an interior act, and how this act introduces height and depth into what we must conceive to have been a flat surface of perception, in such a way that the environment acquires " profile." Now this profile-imparting activity is the—admittedly only half conscious—work of the human Ego. It is this activity which evokes " objects " from the environment, and it is just this which observation has revealed as lacking in the lower animals (Note 25).

It might be supposed therefore that the perception of the *higher* animals would bear a much closer resemblance to that of man. This assumption is largely warranted by everyday experi-

[1] *Philosophy of Spiritual Activity*, p. 171.

ence. It looks very much for instance as if the dog " recognised " his master as one and the same individual. A bird will fly away from the stone that has been aimed at it. The crow " recognises " the gun on the sportsman's shoulder while it views with indifference the farm-labourer's stick. It would seem almost impossible not to admit to the consciousness of the higher animals a genuine shape-world in the human sense.

ABSENCE OF OBJECTS

Careful and systematic observation compel a different conclusion. In the first place we may reflect that the shying of a horse at some object lying in its path depends precisely on the fact that it is *not* recognised as such but rather accepted as pure sense-impression. It is precisely because the animal lacks the " thing " faculty that the impression impinges so sharply, penetrating, as the wise precision of language has it, to " the very marrow of its bones," and making it leap sideways. The calm and indifference with which it commonly regards its field of vision have been rent asunder, and sense qualities dart through, disturbing the beast soul to its foundations. In the case of the bull the thing is more obvious still. The creature runs amok at the sight of *anything* red, without stopping for a moment to enquire about its meaning *qua* object; it is excited down to the depths of its physical being by the colour which " leaps out at it."[1] Here then is something of that chaotic confusion, which in human perception is only to be found *before* its interpenetration by concepts. Only, out of this chaos of unrelated, overlapping particulars the human being fashions, by the application to it of concepts, his articulated and shapely world. The conceptual complement rushes to meet what is merely perceived, and the coincidence of the two is experienced as the act of knowing. It is this putting together of the severed halves of *one* thing which language intuitively depicts in such expressions as " grasping " " comprehending " (comprehendere) and so on.

[1] Rudolf Steiner in a course of lectures given in Penmaenmawr, 1923. " Initiations-Erkenntnis." English translation *Evolution of the World and of Humanity*. London, 1927.

HIGHER ANIMALS BLIND TO THE CONCEPT

It is clear from the instance of the shying horse and the raging bull that animals know nothing of this " grasping " of the other half of reality. The putting together forms no part of their experience. They do not, in the human sense, " get " a thing. They perceive, *but to the concept they are in some sense blind.* It is just as if they lacked an organ for anything that lies beyond a certain limited horizon. They cannot by their own activity unite the percept with its corresponding concept, and what is a " thing " to human beings is to them merely a constituent part among a host of others that go to make up their perceptual totality.[1]

This distinctive absence of " thing-prehension "[2] (which must be carefully differentiated from " thing-perception ") transpires also from the admirable experiments which Wolfgang Köhler conducted with his man-like apes (chimpanzees). These animals astonished all beholders by the way in which, without any sort of guidance, they picked up sticks to help themselves get at some bananas lying outside the cage. This certainly seemed to point to intelligent behaviour on the part of the animals; it seemed to indicate genuine " comprehension." If, however, it had really involved a positive power of surveying the situation as a whole, the following phenomenon, quoted from Köhler's documentation, would be wholly unintelligible. Apparently the use of the stick as a means for reaching the desired objective (fruit) depends on certain definite conditions. If the stick is arranged in such a position that, as long as the animal is looking in the crucial direction, or at any rate is only glancing away from it to a certain limited extent, it remains invisible; and vice versa if a glance in the direction of the stick causes the whole region containing the objective to disappear from the field of vision, then the use of the instrument is in most cases prevented or at any rate delayed in the most striking manner, even though it may have been used many times before.[3] The chimpanzee, then, requires that the tool and the place in which he can use it should both be present

---

[1] Cf. André, 1924, p. 41 ff.
[2] *Dingerfassung.* The translator apologises. Remprehension? "Thing-king"?
[3] Köhler, 1917, p. 36.

within his field of vision in the same moment; otherwise he can make nothing of them. He behaves like an imbecile, forgetting what he wants to do while he turns his head. Even if the tool does happen to lie within his field of vision, it may remain unused if it does not stand out sufficiently from the rest of the environment. Köhler very pertinently observes (p. 83) that the animal cannot " pick out " the object. Thus, if it is looking for a stick, it will altogether ignore the branch of a tree, since that does not stand out enough, and will struggle to pull the bar out of a doorbolt instead. For practical purposes the bolt is attached far more firmly to the door than the branch is to the tree, but optically the former stands out at the first glance. Thus " optical conditions ... may easily vanquish practical considerations " (p. 85). " If one carefully places some implement or other for which the chimpanzee is looking—let us suppose it is a table—with one of its corners fitting into a corner of the room, one will frequently see the chimpanzee pass it by as if it were not there " (p. 87). " There must be a kind of optical fixation, which obstructs the intellectual act of severance as effectively as the stoutest of nails would obstruct the physical act. . . . Given the same external circumstances this optical adhesion is much more easily broken for the adult human being than it is for the chimpanzee, so that the former evidently sees the ' parts ' which the occasion calls for long before the latter " (p. 85).

Could one ask for a clearer illustration of the way in which the human being can " pick out shapes " from a perceptual totality, while the animal is only aware of " what we must conceive to have been a flat surface? " Must not the animal be cut off from that " spark of thought " (Steiner) by which the purely optical unity is broken up into a series of objects?

COMPARISON WITH SLEEPINESS

If we wish to find in our own experience some point from which we can approach this condition intelligently, we may conveniently think of that elusive half-light, in which human beings find themselves when they are over-tired. At such times a man is attacked more violently than usual by purely sensuous qualities, such as smells or individual sounds; he is also much more prone to be terrified by anything darting out unexpectedly at him than he is when wide awake; for in his great weariness he does not clearly take hold of what is happening in front of him. His cognisance of the outer world is, in fact, similar to that of an animal—say, a shying horse. It is only when he " sits up," when he " pulls himself together," that he begins to notice the details, thrusting things away from himself, to a position whence he can observe them with cool detachment. To understand this self-gathering, this off-keeping exactly would be to understand that activity of the Ego which raises human consciousness above the animal level. Man, as long as he remains awake, keeps things at arm's length; but the perception of the higher animals is comparable to the half-light in which he sees things when he is sleepy. A con-ceiving, a " gripping together" in the human sense is nowhere to be found in it.

THE ANIMAL'S SUBSTITUTE FOR THE CONCEPT

The question is, is the animal's perception, though it may not be articulated in a series of objects, after the manner of human perception, completed nevertheless by *something* which corresponds to the concept? For so long as the animal is living in its environment, it does after all behave as if it had something in it like a kind of common sense. We shall only understand the peculiar nature of the animal's substitute for the concept when we have realised that all it can do is to supplement with its ingenious behaviour such situations as already furnish it with the natural conditions. These are the regions in which it is "at home," the circuit of all that biologically " counts " for it. The animal behaves all along like a man with a limited sphere of interests who when it comes to anything beyond them " has no sense." Only certain quite definite situations are relevant for it; within these it feels comfortable, makes various attempts, finds out ways and means; but something else, which for the human onlooker may be just as much " given " and belongs to the same field of observation, is for the animal simply not there.

This is a good place in which to observe the difference precisely.

It might seem at first sight as if the higher animals were really capable of learning something *new*. The things which Köhler noticed in his chimpanzees, such as the use of sticks for reaching things, or to jump with, do make it look as if they had in some sense a new idea. But in actual fact this is not the case. These apparently new acquirements are in reality no more than the metamorphoses of powers absolutely prescribed by kind. Bühler[1] is pertinent enough when he remarks that the apes' use of sticks is grounded in instinct, the stick taking on, since they are tree-haunting animals, the functional value of a branch. The monkeys merely carry their activity a little farther within the (to them) familiar branch department. It is easy enough to see that a tree-haunting animal must be continually meeting situations in which it is obliged to operate with branches exactly as if they were sticks, except that the former grow. The monkey using a stick, then, does not have any new "idea."

This brings us to the heart of the difference between the animal's world and man's. It is only man who is in a position to keep on adding more to those parts of the world which he has comprehended, proceeding from these to yet other parts, and so on *ad infinitum*. Instead of the biologically significant he grasps the factually "important," however irrelevant the particular facts may be to his own biological welfare. The human being's interest—in the literal sense his "being amongst" things—is applied to the world for its own sake. He finds thoughts lighting up in him in quite new connections. To meet the fresh percept, there arises in him the new concept which belongs to it: whereas within the animal's narrow horizon there are no possibilities beyond awareness and activity. It lacks the necessary organs for anything that lies outside. It is inserted into that sector of the world for which its physical organs are adapted, and is not in a position to penetrate with insight into anything lying at a distance. So that even in situations in which it behaves sensibly, it does not do so with discernment. Unlike a human being, it cannot consciously unite a percept with the corresponding concept, and

[1] Quoted André, 1924, p. 38.

thus penetrate, through knowledge and deed, to " reality pure and simple."

That even the man-like apes cannot do this is clear once again from Köhler's exposition. Again and again in the course of his experiments he was brought up against boundaries, set by the essential nature of the beasts themselves, at which they began to give up; they were held fast within their own horizon, and Köhler, with his marvellously acute sense, ended by taking the hint and confining his experiments to this circle. "One is continuously being astonished," he writes, "often positively annoyed, to discover that, no matter how clever and adaptable an animal of this kind may be, every attempt to transform its innate biological qualities slips ineffectually away."

It is true then that the animal " has " a certain circle of " substitute concepts," but this is all it has, and even this it does not have in the human way. Instead of waiting to be united by a free act to the percepts, they are attached to and can only be made to function by an outer event. The animals' restricted circle of concepts is born with them, and it needs no exceptional acuteness to understand that in their case the relation of the conceptual element to its correlative sense-datum is one of mere reaction. Moreover, since the circle of innate concepts is determined according to the animal's kind, it is also of course closely connected with its physical organisation. Man, in his thinking, calls down the concepts from a spiritual world (where even concepts are alive), in order to accomplish freely, and out of his will to penetrate the world-system, their synthesis with the percepts. This is just what the animal does not do. For the animal the conceptual element is located in that dim-lit region beneath the level of the waking life, in which the consciousness proper to organic activity resides.[1] There it waits, as it were, for the opportunity to encounter its sensuous complement, with which it immediately fuses. If this occurrence were actually experienced as a putting together and as a mutual correspondence, it would be an act of knowledge. Instead of that, what might have become knowledge takes a different direction, discharging itself in the form of physical activity.

[1] Compare with the previous chapter: Wisdom influencing the organism.

THE UNKNOWN COMPELS, THE KNOWN LEAVES FREE

And so we arrive by a different route at the same conclusion, that the duality of knowing and doing is a purely human affair, and one which even the higher animals have to forgo. The coincidence of the sense-datum with its physically impounded conceptual complement, unnoticed by the animal, reduces itself to mere " behaviour." The animal huddles together our knowing and doing, retaining as a substitute for the first the instinctive operation of the innate concept and, as a substitute for the second, the translation of an unknown something into unreflecting physical activity. Thus, that active presence which in the case of a human being is inserted between the two poles—the self conscious Ego—is crowded out. It is this Ego which is able to apprehend and hold fast the hidden relations between things; and it is out of this same holding fast that the will to act is born. But in the case of man, while this will is indeed built in the first place on perception, it does not simply result from perception as its effect. On the contrary it expresses the beginning of something quite new within the world-process. Facts are created by it for the first time.[1] With the animal an instinctive and congenitally determined knowledge—which is a non-knowledge *of things*—passes over imperceptibly into congenitally determined behaviour. An experience, which is not experienced, turns into a deed which is not done. Man first apprehends and knows, *and then* applies himself to do. It is his condign privilege to keep with his Ego watch and ward on the all-important bridge from the experience to the deed.

ANIMALS HAVE NO " BRIDGE-WARD "

Once this connection is realised between the instinctive non-comprehension of the animals and their obscure compulsion to act, it becomes clear in what way their soul life is excluded from that status which, as sentinels on the bridge, the souls of human beings occupy. The animal soul cannot project itself into the actual being of another individual, feel this other's experience as its own and direct its behaviour accordingly. It is obliged instead

[1] *Philosophy of Spiritual Activity*, p. 163

to content itself with all that which—dreamlike, misty, and yet powerfully effective—flows into it from the depths of the physical being of its whole group; and it must keep to such organic impulses as, without grasping them at all in free intuition, it shares with the group as a whole, and which come to light for us in the form of collective behaviour. Group behaviour belongs to the members of a tribe, a herd, a stock, and arises from their physical nature, almost as unmistakably as the nesting odour and the universal chorus of song (Note 26). Right up to the man-like forms, we find this instinctive community, which we cannot help admiring, though its significance has been too long obscured by the anthropomorphic legend of a " struggle for existence."

THE BOUNDS OF THE INDIVIDUAL TRANSCENDED

Every animal activity, since it proceeds from the collective Ego of the whole group, and not from a personal Ego, as with man, can always cross to and fro the frontiers of individual existence. The stampede of a grazing herd, changing directions several times in its flight, is an impressive example of this; and in the war[1] many a member of a mounted unit experienced for himself the way in which all the horses of a squadron, terrified by the shells suddenly falling near them, would carry out a sudden turn, just as if the proper word of command had been spoken without their riders hearing it. An instantaneous resolve, issuing from the unknown, took hold of a hundred animals at once—no time for one to be first and the others to copy it.

This transcending of the frontiers of the individual was also most pronounced in the case of Köhler's chimpanzees. The psychologist used to give his beasts empty wooden boxes of different sizes, with the help of which they were able, by piling one on top of the other, to get at some coveted object such as a banana suspended from the roof of the cage. Now if a second chimpanzee watches the efforts of one of his mates, " he often cannot help putting a swift hand to one of the boxes that looks like falling, to steady it, just as the first beast is making a crucial and dangerous exertion " (p. 132). In one of the photographs taken by Köhler

[1] 1914-19.

one such beast may be clearly seen holding on to a wobbling box —part of the artless tower—while the other up above reaches its "objective." Nor does the assistant look up in an envious way while this is going on; it seems rather to be wholly lost in its obscure necessity of holding on tight. Obviously it does not in the least know, or at any rate knows no longer, whom it is helping and why it is helping. "Something exactly like compulsion," as Köhler himself expresses it (p. 133), has taken hold of it.

A GLANCE AT THE OUTER FORM

This kind of unfree overlapping of the will impulse into other individuals is an infallible indication that the animal Ego really does not, like the human, penetrate into a single personality. This brings us once more to a consideration of the animal organism and once more we find that we can express the bodily difference between man and animal as follows: the animal's dropped head prevents the world from revealing itself as a duality, the conquest of which might beget knowledge; while its limbs, because they are not fully given over to the earth, do not permit of a free and individual resolve being conveyed through them into universal reality. Whereas the human body, with its upright posture wrenching the head clear of gravitation, allows the human Ego to penetrate to a knowledge of the existence of things, and assures freedom to human action. The prophetic powers built wisely, when they built a way for the human entelechy into the world.

WHAT MAN GAINS

At the end of this chapter let us point out once more what powers man gains through his different relation (as compared with the animals) to his organism.

What is the very essence of the human being's capacity to perceive things? What is the *sine qua non* of his knowledge of a world outside him? If we are not content, by way of explanation, with the comfortable old idea of the eye making a picture like a camera-obscura, while the optic nerve conducts this " excitation " to the brain, where it is supposed to be transformed into sensation (a misleading notion which Rudolf Steiner spent more than a generation in combating[1])—if, instead of this, we hold fast to the unbiased evidence of experience itself, we arrive at unexpected results.

Our sense-organs can only fulfil their function as long as the organic processes which play through them remain outside consciousness. Any disturbance of the lens or corpus vitreum, any rush of blood to the retina, disturbs perception. Once the nerve is inflamed, perception ceases altogether. The proper activity within the organ itself must remain unnoticed. Or, if not, then, as in the case of a roaring in the ears, we hear nothing. We can feel nothing with an itching skin, taste nothing with a sore tongue. So far then from the photochemical decomposition of the visual purple in the eye or the electric currents in the nerve elements having the task of creating sensations, they have a totally different function. It is their business to leave the way clear for the Ego out into the world—not to hinder it from reaching " things." The eye, said Rudolf Steiner, sacrifices itself, as it were, in becoming perfectly transparent, and it is this that makes seeing possible. It is reasonable to suppose that the same thing applies to the nerve and cerebral processes—they must likewise " recede " when anything is to be thought or known. According to the *Philosophy of Spiritual Activity* (p. 148), " the human organism suspends its own activity, it yields ground; and the ground thus set free is occupied by thought." Another self-sacrifice of the organic, in order to leave the way open into the world!

If on the other hand the organic processes press forward into consciousness, the Ego is torn from its communion with the outer world. Sufferers from migraine are acquainted with a painful illustration of this. Absorbed by its experience of processes at the centre, the head aches and participation in, or knowledge of, outward events (especially in a violent attack) is prevented.

Now we have already described the way in which, with animals,

---

[1] *Philosophy of Spiritual Activity*, pp. 70-71, and Einleitungen zu *Goethes Naturw. Schriften* III, p. 10, *et seq.* English translation: *Goethe the Scientist*, New York, 1950. Ch. III.

the organic processes of the body, including those within the sense-organs, send up tendrils, as it were, into consciousness. This is why the animal cannot even in normal circumstances apply itself wholly to the outer world; it is constantly preoccupied; for it is following out its own sense-processes. Consequently it cannot gather itself together out of the half light in which it dwells, any more than the human being with a headache can win a struggle against the blood hammering in his head (Note 27). In this way we may realise for ourselves the way in which our humanity depends on a transparency of the upper part of the organism.

THE METAMORPHOSIS OF PAIN

Rudolf Steiner has even described, from supersensible investigation, how the human organ of vision formed itself in primeval epochs out of a kind of wound inflicted by the forces of light, a painful burning sensation, as it were. Vision only became possible later, as these painful wounds developed into transparent organs of light.[1] So that we might truly say: *the human being's apprehension of the world is brought about through the subduing of pain.*

Turning now to human activity, we find that this too was produced by the conquest of organic processes which seek to maintain control of the will. With animals it is a physical stress that impels to action, not only in regard to food and reproduction, but in all their doings; hence it was said above that the animal "enjoys" in all its activity a kind of relaxation, a liberation from the pressure of congested organic forces. But human activity originates precisely in the damming back of this pleasure that is felt in organic activity and in a motivation into which such pleasure no longer plays. In free intuition therefore " the involuntary activity of the human organism is paralysed, pressed back, and the spiritual activity of the idea-filled will takes its place."[2]

Among human beings this secret working upward of organic processes into the will only occurs under pathological conditions.

[1] See the Lecture-Cycle *Das Karma des Materialismus* (1916), Lecture 3.
[2] *Philosophy of Spiritual Activity*, p. 211.

It is the condition of hysteria. In hysteria something uncomprehended, heard rumbling on beneath the threshold of consciousness, impels a man to uncontrolled actions; and these actions produce in him an ill-understood and hardly noticed pleasure of a positively organic nature.[1] Yet this type of concept that rumbles on, as it were, in the organic realm without ever penetrating to consciousness is thoroughly characteristic of the instinctive behaviour of animals.

METAMORPHOSIS OF PLEASURE

Once again it is Rudolf Steiner's spiritual investigation which confirms the fact that action arising out of a sense of physical well-being was in olden days an attribute of humanity. He says[2] of Lemurian man (see Chapter II, p. 58), "Everything in his environment as well as the pictures in his soul stimulated him to activity, to movement. If his activity could vent itself in an unhampered way, he felt a sense of well-being; but if it was repressed in any direction, he was uncomfortable and ill at ease." And so one might say: human activity is brought about through the subduing of organic pleasure.

In its perception the animal is involved in a process which, for men, would betoken bodily pain; its activity is of a kind which brings the processes of tension in the organism to a pleasurable issue. Human Being on the other hand is a healing from the painful influences of the outer world. On the perception side it is a vanquishing of pain; and on the willing side it is self-liberation from the ease that is bound up with organic activity—a vanquishing of pleasure.

The head, set free from pain, rises erect and gives itself to the apprehension of the world, as it fashions ideas corresponding to reality. The limbs, quit of organic pleasure, give themselves to earth; and, placing themselves in the service of what is contemplated in the spirit, they make it come true in space.

[1] *Lectures to Teachers*, London, 1948.
[2] *Lucifer Gnosis*, No. 19-23, p. 32, first published 1905. English translation *Atlantis and Lemuria*, London, 1911, and *Cosmic Memory: Pre-History of Earth and Man*, New York, 1959.

THE HIGHER ANIMALS	MAN
(Warmblooded)	*Pole of Perception*
	(Painlessly surrendered to
Percept working downward	*the outside world)*
into physical process	

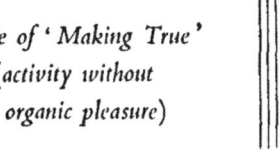

	Human Ego
	(free)
	Pole of 'Making True'
Physical occurrences working	*(activity without*
upward into consciousness	*organic pleasure)*

Diagram 16.
Outline showing the difference of consciousness in man and in the higher animals.

Between perception on the one side and " making true "[1] on the other human nature unfolds itself. But pain and pleasure, released from those bonds which fetter them to the organs of the body, may become organs of the soul, may become helpers of a human heart which learns to avail itself of them in freedom.

And so we come to know man as in very truth a Being made whole from the sickness of becoming animal.

[1] Translator's Note.—It is impossible to convey the German antithesis of *Wahrnehmen* (literally, to take for true) = to perceive and *Wahrmachen* = to make true.

PART V

DESTINY

The biography of one man corresponds to the history of a whole species in the animal kingdom.
—RUDOLF STEINER, 1904.

THE ATTEMPT TO describe clearly the essential difference between man's being and the animal's, as seen first from outwards and, then from within, attains its full meaning and balance when it is directed to that plane where opposition between outer and inner ceases. That is the plane of destiny. Here is the heart of their essential difference. All the lines of the picture must converge here.

People of the present day, however, have only an inadequate conception of destiny. They think of hidden laws ruling behind the events which happen to one people or to another and describe these as natural laws. These are represented as working from outside: each single event " obeys " these laws just as in lifeless Nature each particular case depends on the ruling law. Destiny becomes a threatening or even a crushing power. Some persuade themselves that this power is the working of the stars, and thus seek a highly dubious renewal of the old astrology.

Any such view of fate, as a power controlled from outside man, is a great obstacle to a true comprehension of the riddle of destiny. Whoever would advance to this, must labour to get through to an entirely new conception of fate. If he cannot abandon all comparisons with processes in Nature, it would be better that he should entirely deny that there is any connection whatever between the various happenings in a man's life. The half truth is more dangerous than a complete lie.

THE SHAPING OF DESTINY

Rudolf Steiner in his book (1904), when he saw this difficulty, felt obliged to give a quite independent elementary description of the moulding of human destiny. None of the cloudy ideas in current use could harmonise with it. That mysterious happening, the birth of destiny, must first be unveiled before the inward eye of the investigator.

The reader of these last chapters can start from the fact (see p. 82) that man's evolution has suffered an extraordinary retardation as compared to the animal's. A surprising difference of rhythm arises because of the diversified activity in the inner nature. In the animal world, the age of puberty signifies the conclusion of the development of soul and body, for man this is the time at which he begins to make real progress towards becoming fully man. The animal has reached maturity when it can propagate its kind; this is the period when man is only just starting on the development which distinguishes him. His body has to remain "young" so that he himself may grow to be a man. The animal is already "old" when it reaches puberty and would show an outer ageing also, did not the reproductive forces rejuvenate it.[1]

Man, after attaining a physical development capable of propagating the race, can still remain young; but he differs entirely from the animal in that this is not simply due to his reproductive organs. The entire earthly being remains plastic, the outer body is still rather like a child's, right on to the twentieth year. And gradually this plasticity passes over into soul and spiritual capacities. Among animals it has long been concentrated upon propagation as its central function.

From Rudolf Steiner we know what it is that is served by the retention of plasticity, and what has to work down into this plastic organisation of soul and body; it is the human entelechy itself which is at work on its own sheaths. The Ego which is clothing itself in a body needs this stemming back of development during the embryonal and the adolescent periods; without it, the hardening process must lead on to animal formation: further, when the youthfulness of the body is coming to an end,

[1] See *Lectures to Teachers* (1921), Ch. 1, Sec. 3.

it is the Ego that maintains in its place the soul-spiritual plasticity. Once again the human Ego appears in its characteristic role as that restraining and delaying power which, in the first instance, it revealed itself to be in the evolution of the race (cf. Chapter II). Now, in an individual life, the Ego has, as its first task, to keep the organisation plastic, to restrain it from hardening prematurely before the time has come in which it can enter the world of earth: now only does it descend into these " sheaths ": now, knowledge and action begin, and now, instruments which had been prepared beforehand are brought into use. Before this the Ego was at the service of the sheaths and guided their unfolding, now they must serve the Ego as mediators between it and the world: they provide the Ego with information about its surroundings, and in them (the bodily sheaths) the Ego intervenes in the world through its actions.

Thus we are on the track of the activity of the human Ego when we observe man unfolding his powers. Everything that really plays a part in his development may be considered to be connected with the Ego itself. We usually take account of the circle that belongs to a man in too narrow a sense. Often the physical body is looked at in this connection quite as though it could exist without its immediate or remoter environment, and yet these are as much part of the man as the particles of his body. Further: not only do his surroundings bear the stamp of belonging to a particular Ego, but also, for closer observation, every relationship with other people belongs to it, each lasting or fleeting connection and association in life; in the wider sense indeed, each experience belongs to a mysterious whole which is, for any individual man, just as significant as bodily peculiarity. This is most clearly seen from examples of notable personalities. We see that to the physical features of the face or countenance others are added, from the biography, without which the former do not constitute a true and complete whole.

In the life of a great man how convincingly each step taken, each utterance, each action fits into the whole configuration of the man. The smallest and the most fleeting things have significance, and resemble no chance pebble on the road, but rather some fragment of a great and marvellous mosaic: it only needs reflection to see the pattern of the whole, and by taking a few steps backwards as it were, to readjust the details which could not be seen at too close quarters. As an example of this, every scrap of writing from the hand of Goethe, every remark he dropped, every little conversation, every meeting, all bear the stamp of his personality. After all why should we doubt that the meeting with another human being may be dependent on our own individuality merely because, out of our usual experience, we cannot imagine that any such passing " encounter " is caused by our own Ego? Do we really understand any better how we ourselves, all unconsciously, take substances from the outer world into our own organisation and incorporate them in our bodily processes? The chemistry of this is itself an individual process; revealed in specific attractions and repulsions, it bears the stamp of the Ego as surely as the countenance does. Corresponding to the individual alchemy of the body, should there not be an individual capacity for assimilating *events* to our particular path of life? Does not our own Ego perhaps interweave itself with the encounters of its life in time, much as it affects the matter proper for the body in space?

With this our survey has advanced to the heart of the problem of destiny; that close connection between an experience and the person who has the experience which natural law cannot explain. Their actual inter-dependence is often so strikingly clear to the average observer that it has become proverbial. We see one person with the " lucky touch " in everything he does. Another is a hopelessly " unlucky beggar "; and when somehow or other in the ordinary course ill luck again befalls him, again we say, " it couldn't have happened to anyone else! " here we are on the track of the inter-dependence of his inner self and all that happens to him. Inner and outer are related to one another; how this happens is no doubt a mystery, but it is none the less a patent fact. The closer we look into it the more clearly we see that what is true of the man who " never has any luck " is equally true of the man " born with a silver spoon in his mouth," and indeed, of every man. No one can doubt that in every man's life things happen which are so entirely characteristic of him that his own face might be looking out of them. It is indisputable that there are very many facts in an individual life about which it can fairly be said that this " could only have happened to him " and that " it is just like him."

AN EXAMPLE

When we have grasped this close relationship which exists in certain given instances and which cannot at first be explained, something more follows. In making observations of this kind it is well to begin, not with one's self, but with an historical personality; for in this way we ensure objectivity. Let us recall the famous experience of Galileo in the Cathedral at Pisa. We can picture him sitting in the church, watching the great lamp as it swung slowly above his head. In his soul perhaps was weariness, or some such feeling, certainly no kind of definite intention. He observed the slow swing to and fro and at the same time noticed the rhythm in his blood; it then occurred to him to compare the two, and understanding lit up in him: the law according to which the lamp was swinging, the law of the pendulum, was discovered. Thus Galileo, letting his mind wander at Mass, became the discoverer of a law without which our present-day physics could not have evolved. One is quite definitely obliged to acknowledge that it needed a Galileo to have an experience *of this kind* in church: for other men before him had of course been in like circumstances without doing anything at all for the advancement of knowledge. Not only was the experience itself characteristic of Galileo, but also *its outcome*. From contact with this occurrence there issued a definite result which bore the mark of his individuality yet more clearly than the occurrence itself.

THE ROLE PLAYED BY INDIVIDUALITY

This has taken us a step further. We begin to understand that the human Ego looks out for events and encounters and chooses among them. Out of a survey infinitely more extended than that which is open to the ordinary consciousness it has to shape the course of life. In guiding the course of experiences it must have had regard to their issues. It has had to seek out particular tasks, tasks which follow from its very nature, which are not laid upon it from outside, but to which it has itself summoned itself.

When we have gained such a view-point, we can see the human Ego before us in the might of its real presence. The Ego, for the earlier psychology, was nothing more than a collective name for all these vague subjective shadowy experiences. It is now coming into its own as a being, an active being, which not only takes masterful hold of the course of bodily development, but also inspires the most vital events of an earth-life. Right back beyond the beginning of this life our sight is directed to a preceding period during which all that is prepared which later comes together as the unity of an individual human existence. Over all these preparatory conditions the Ego itself must have watched, waiting for and deciding upon the hour when it must connect itself with the earth.

If conscientious thought is obliged to admit all this as a possibility, to Rudolf Steiner's spiritual vision it was an actual process. And it was a process which he could describe. Out of his own direct personal vision he reveals the pre-natal conditions of the human individual, the watching over the coming into existence of the germ of the body, and the waiting for the right constellation for its entry into earth life. Without this pre-existent kernel of his true being which influences all that happens to the body, and overshadows the course of destiny, and which awakens to self-consciousness when the child learns to say " I," without this, man is like a sheath with no content, a shell without a kernel, a garment without a wearer.

The responsible part played by this Ego-being must be kept constantly in view when we seek to comprehend progress on the path of an individual life, from the earliest stages of the development of the embryo to the death of the body, from the earliest events of childhood (by no means conscious experiences as yet) to the experiences of the mature and ageing man. Upon the germ of the body which it takes over from the body's ancestors, the Ego stamps the individual features of its own character (see Chapter III, p. 95). It imprints its seal also upon the path of life itself, until, emerging from the series of outer events of life, it takes on a living form, whence the spiritual countenance of a man looks at the beholder. When the Ego, overshadowing a body and a life, is regarded as one whole, we can comprehend what each experience must mean for the Ego and its bearer, the body. We see the human Ego entering the earth world, confronting it with senses and with thought, by its activity taking hold of the earthly world and altering its conditions. Now Rudolf Steiner leads the beholder to the point where he can watch the most intimate growth of a life and its results. He directs the gaze to the mystery of memory. This has in the first place the capacity for preserving the impression of a past experience. And by means of this quality—to quote from Rudolf Steiner[1]—" the soul makes the outer world so into its own inner world that it can then retain the latter in the memory, as a recollection—and independent of the impressions acquired, can lead therewith a life of its own." It becomes clear that every impression, as it flows into the totality of soul-content, must be altered by the individuality it meets, by whom so to speak it is coloured. It does not pass into the memory unchanged—and it can attain a renewal of its shaping power only if the Ego works upon it again. When it has sunk right down into the unconscious regions of the organisation, it is no longer a memory-trace, but because it has been worked upon by the Ego (the inspiring power) it has become—together with other formless, that is to say, transformed experiences—the real individual possession of the soul. From now on it is active solely as *capacity*; this is the form in which it continues to work on in the permanent quality of the individuality.

EXPERIENCES AND CAPACITIES

When these submerged experiences are completely transformed and permeated by the Ego, we get something essentially human. What a single experience can mean when it passes down into this storehouse depends upon all that had previously been stored there. Something previously experienced opens the eyes to things

[1] *Theosophy*, English Edition 1954, p. 62.

which would not otherwise be noticed. Ideas already grasped open the way to new experiences. Every joy accepted moulds the inner being. Perhaps it is the suffering of pain that takes the firmest hold, not indeed to destroy, but rather to mould more vigorously and to help on to final maturity.

Thus it can be said, that for the soul and spiritual development of man *the sinking down*, the submergence of the single experiences plays the most significant part. Let us consider for example, as Rudolf Steiner has described it,[1] what happened when as a child one learned to read and write. How difficult everything seemed, how incomprehensible and quite unattainable the powers of full grown people: and what must it not all have cost in minor experiences, progress and failure, encouragement and blame, hope and disappointment, insight and error! Meanwhile all the details, however vital they may have been when they took place, have disappeared. " But no one could read or write if he had not had these experiences." Thus man's experiences, even though really great, could not alter him, if he could not forget them. Personalities which are continually repeating their experiences in their souls, cease to grow.[2] Those who have suffered an injustice and keep it always in mind, troubling everyone else with it, miss the blessing of mature experience. Vindictiveness, bearing a grudge, makes a man ill. To be able to forget encourages and renews him.

All these transformations are signs of the activity of a human Ego. It is the Ego which moulds experiences and their consequences. It distils enrichment out of destiny: it sets an active self-transformation over against the passive " Oh, let it be," and redeems all past suffering to become the foundation of future deeds. Thus the Ego prepares the new destiny. Man constructs in learning; he creates in forgetting; both faculties are his ministers.

★ ★ ★

[1] *Theosophy*, pp. 65-66.
[2] The case of hysteria on p. 107 should be compared with this. The awful thing is that the sick person cannot forget. She can therefore develop no further power out of her sufferings.

A continuous stream is flowing out from the world into experiences, from experiences into capacity, and from capacity again, as deeds, flowing back into the world. Everywhere the Ego is active, *prescient and forewarning* amid the concourse of experiences, *transforming* where capacity is being formed, *creating* where the impulse to activity is pouring into the outer world. Expectant and discriminating, forgetting and perfecting freely, with open hand giving back to the world the gifts received, thus it is that the Ego stands in the current of life's events. Because a man bears this and no other Ego he has particular experiences, out of which he develops particular capacities (or incapacities) and from which certain deeds—and misdeeds—issue.

As we watch this stream flowing out from the world to man and from man again back to the world, the question of necessity arises: why is it that this stream returns into itself? Out of past experience it flows into the present quality of the human soul and out of this into deeds and their effects, and from these on into the future. What if this future will at some time be the past? How if the deeds performed are also, further on ahead, to remain bound up with the Ego? Might not the Ego which was overlooking, guarding and guiding the whole of the stream's course, determine that from all this, in the far future, new experiences should develop?

But let us keep this thought in the background for a moment: past experiences in man's life flow together and unite, and this past clearly belongs to him; he retains it within himself: but his deeds separate themselves from him, they enter upon the scene, disperse themselves in the places where he lived, make their consequences felt, and are all the time continually receding further in time and space. There the stream seems to disappear, to be absorbed by the wide world. But why should those deeds and their consequences belong any the less to the Ego, because, physically considered, they appear to be getting further away from it? Certainly, they belong to it: they carry the impress of its individuality and the idea is tenable that they retain in themselves the urge to approach the Ego again, this time from outside. Steiner himself puts this thought before the unprejudiced observer who must test its consistency[1] for himself.

The thing that will bring it home to the mind most effectively is the realisation that man, when he enters upon life, *brings capacities with him.* That capacities are built up out of experiences can be directly observed by everyone. Only one possibility therefore remains—to relate all the aptitudes unfolded by a soul to the experiences of an earlier earthly existence.[2] Thus both the spiritual stature and the spiritual countenance (its very cast) appear before us on the basis of previous life-histories. Goethe's spiritual stature—which stamps his genius on all his experiences and deeds—appears as growing out of a series of earlier stages of existence, just as the bodily form was laid down during the series of the generations.

REINCARNATION

Now that which is here presented to the reader as an idea, for the seer with spiritual vision becomes a visible reality. Rudolf Steiner has described in detail the evolutionary path of life of certain great individualities.[3] His descriptions abound in the most arresting inner connections. They disclose a plenitude of drama and tragedy in the evolution alike of individual man and of humanity as a whole. This is one of the domains in which this teacher stands alone, in the width and greatness of his survey.

HUMAN DESTINY

The foregoing enquiry, however, is not concerned with the destiny of single persons. It is only able to indicate what is the essentially *human quality* in the facts brought to light by Rudolf Steiner: The knowledge that the human Ego is substantially indestructible, and continually finds itself anew on the earth in succeeding ages of time, solves the riddle of destiny, and its relation to the personality.

The individuality passes through a series of incarnations in the different epochs of human history; in each life it appears as

---

[1] *Theosophy*, p. 64   [2] Ibid., pp. 72-73
[3] *Karmic Relationships*. 4 Vols. London, 1955-7

another personality, as the member of another race, of another family with new surroundings and new tasks, which it has sought out and chosen for itself. In ever new metamorphoses, prepared during long intermediate spiritual intervals, there arises a series of higher and higher stages of perfection, within which the individuality, always with new means and against new hindrances, wrestles its way through to consciousness of its own true being. From life to life the relations to other men are woven; what has been left behind is once more taken up; what was broken off is united again; what was failing can once again be made good. Each time the scene is chosen anew: the surroundings, the time and the race; each time a body, capable of preserving its plasticity, is formed anew. Experiences are sought out which are moulded into capacities; activities are undertaken which are in themselves an expression of the living being concerned; and they continue working, so that later—like changing constellations—he is to meet them again, having meanwhile changed himself. Thus destiny, *human* destiny, becomes not an overpowering fate and oppression, but a freely sought task, an opportunity to make good, a spur to perfecting the human being. Human destiny is individual destiny; it is the outcome of decisions made before birth and continually grasped anew. The way in which, of his own impulse, the individuality descends again and again to earth life, and inserts himself into the path of the whole of human evolution, marks the particular mission and role apportioned to man in contrast to other creatures. We have here no chain of blind chance, nor of fate; but a recurring plunge, each time more resolute, into the light of the freedom allotted to man. It is the destiny of man to ripen towards self-determination.

ANIMAL DESTINY

The relation of the animal to its destiny is entirely different. When we try to understand this, we find ourselves in another world altogether. But here again Rudolf Steiner has given decisive indications.

In the life of the animal, from conception to sexual maturity, an uninterrupted development takes place; it begins with the fertilisation of the germ and ends with sexual maturity. The apes were cited in illustration. How striking is the difference between their uninterrupted growth and the delayed development of the human frame. There is a positive haste towards culmination; in tremendous strides the organisation advances to its perfection. It retains no plasticity. The body hardens and all formative shaping forces pass on and over to the embryo states of the succeeding generation. This body, which we saw previously (Chapter I) to be completely developed for its purpose, to be perfectly formed as a tool, an actual tool, is interwoven with its surroundings in a way quite other than that in which man lives in his environment. The environment of any species of animals, properly speaking, belongs to its bodily organisation; the two are complementary or, as we say, " dovetailed " into one another down to the most intimate details. This is just the miracle which led to so much cudgelling of the brains by the investigators of previous days, and which Darwin believed he had solved by his teaching about " adaptation to environment "—the extraordinarily subtle interplay between the organisation of an animal and the scene of its life. In so startling a manner is the animal body stamped, so to speak, into the field of its existence, that other conditions are simply unimaginable. Thus it comes about that when we are giving, say, a detailed description of the bodily structure of the butterfly, we digress unawares into a description of the surrounding world. The very line with which we trace the shape of its body takes us out and beyond into the details of its surrounding world, only to return once more from these surroundings to the creature's own body and organs. The creature lives in this scene of action, all mobility, yet like a prisoner. Further, this impression of being in confinement is heightened when we consider that the animal—as was shown in Chapter IV—does not really perceive its life-region as an objective world which it can confront, but as a visionary world dimly seen and which keeps its soul in captivity. The experiences of the animals are entirely subordinated to powerful impressions received from their surrounding world, mixed up, as these are, with impressions of their own inner organism. The cleavage between outer and inner world is a *human* experience;

the animal knows nothing of it. And just as bodily organisation and scene of action are firmly interconnected, so also are the experiences of any individual animal prevented from getting beyond a narrow circle with a particular and given content. The stiffness of the body corresponds to a rigidity of soul experience. The stiffness of the body permits of no Ego taking up its abode therein; the animal Ego cannot step forth on to the "physical plane"; it has to remain in the soul world, thence to guide all the animals belonging to that particular group or kind. It only dreams of all that happens physically to the single beasts. It is incapable of making free decisions, and the impulses which it imparts are common to all.

If we add to this what has just been said about *human* destiny, we recognise that here the preliminary conditions which distinguish human development are lacking. These are an organisation remaining open, so that the Ego (with its formative power) can work into it, and a capacity to let experiences sink in and, with the guidance of the Ego, to develop new abilities (Note 28). It is precisely the complementary nature of organisation and environment which hinders the kind of experiences best able to forward evolution. Those all-important encounters and events which carry a *man* further arise—as everyone knows—precisely when he confronts a new situation where, for instance, a question is put to him for the first time, or he suddenly finds himself in a dilemma—in brief, the great experiences which form him, arise out of the discontinuity and disharmony between man and world. Particularly in great personalities, we see how much of their beauty and excellence is really due to trials suffered earlier at the hands of the world. Beauty—as many have recognised—is pain suffered and transformed. Because the animal is adapted to its environment, it is denied the possibility of developing inward maturity and greatness. As an individual creature it cannot grow beyond the limits of its kind; and again, at death, it falls back with its capacities into the group Ego, from which its soul was something like an offshoot or a patrol sent out on reconnaissance.

The animal has no individual destiny. It only shares the common destiny of the group, just as it shares its organisation and its environment with those of its kind. Its destiny is predetermined, its individual being only emerges for its life-time out of the element in which it is fast-bound, out of the stream of heredity and descent. From this it receives its body—and unlike man, who, with the Ego present in the body, is able to stamp his individuality upon it, the animal, lacking this presence, gives out its formative life-forces for the development of its descendants. By way of these hereditary-forces, the animal receives its impulses—impulses to gregarious rather than to individual behaviour—and on the return flow of the stream of heredity, the animal's experiences, typical of the group (though only dreamily conscious and incapable of transformation), are carried back to the group-Ego. In the things that happen to him man can recognise his own Ego.[1] But in the things that happen to an animal the species can be recognised. The characteristic features of its destiny are decided inflexibly and indeed as firmly as its bodily physiognomy.

ANIMAL FORM

If after these considerations we direct a sensitive observation to the forms of animals, we receive the most unexpected impressions. We begin to comprehend the mysterious attraction that the sight of an animal has for any normal person. It is always strangely moving to look, in an animal, at that which in man is called the face. It is difficult to give expression to this experience. It is as though something *veiled* were living behind the physiognomy. Something that *craves* to shine through but is withheld by the body's rigidity! This impression becomes positively grotesque and horrible in the case of an insect. Looking at the head of a wasp or a butterfly, perhaps through the magnifying glass, we cannot help almost shuddering. The merciless rigidity and hardness of the casing, out of which the eye, an immobile point, its surfaces walled in, stays there lidless and ever open; that fearful leverwork of the parts of the mouth working mechanically; the hurried jerks and cramped groping of the proboscis; the antennæ, always trembling, and yet not looking truly " alive " —it is as if one saw a ghost, a phantom suspended by invisible

[1] *Theosophy*, p. 82.

threads, that pretends to be alive but in reality is only a moving mechanism. Our darling butterfly (under the magnifying glass) is a masked and hideous creature.

A monstrous mask is weaving around the animal head. Even the creatures that live nearest to us, our domestic animals, look at us with an expression which leaves the observer hesitating between compassion and fear; and which of us has not looked at a horse with more and more perplexity, as, behind the unspeakable expression of its face, we tried to communicate with the creature itself, that would fain speak out of this face but cannot? Does not this set form, so like a face, hang like a veil in front of the real being hidden there? In the higher animals the head is a sort of mask through which the creature's real countenance cannot look. The impression grows, until it becomes almost unbearable, when a dog, for instance, instead of laughing like a man who can give expression to his inner self, bares its teeth and—grins. Only blind infatuation could say that the dog laughs; an unprejudiced and unsentimental person can hardly bear the sight of it. There is scarcely anything on the earth outside the human kingdom that can arouse such strong feeling as an animal countenance. The whole fate of a being that is not free is written across it.

In looking at the animal physiognomy (and the same holds good for the whole form), we can divine all its remoteness from the Ego—barred as that must ever be from such a body—and we feel at the same time the whole guiltless tragedy of the creature banished from the possibility of free experience and free activity.

ANIMAL VOICES

With this we begin to comprehend the strangely moving impressions made upon us by animal calls and cries. The bird chorus so welcome to us in its selflessness and innocence does no violence to human sensitiveness. But still more impressive is the sound of a herd of cattle bellowing in the night, or the dog's yelp bringing answering barks from the distance. It seems as if these tones resound out of profound subterranean depths in the body! It is not the ox that bellows, not the dog that barks—a bellowing comes from the ox, a barking from the dog—*through* the dog. From the land of dreams it pours into the world of man, and stays there, isolated and strange, among human sounds.

But the human voice sounds out of the man himself, out of his breast, it proclaims his own character, his feeling, his willing. The voice of an animal is wrung out of it as though by a nightmare; not produced in the free course of the breath—it is full of the destiny that is suffered but not understood! An unredeemed being is striving for expression in it. It sighs even when the creature seems to exult and sing.

ANIMAL PLAY

And finally, if we are watching young animals at play, then in their play this impression of something of a human likeness is redoubled. The puppies tireless romping, the impudent drolleries of kittens and lion cubs, the obstinate awkwardness of little kids, watching these brings home to us anew our close connection with these creatures. But at the same time there comes a realisation of their distance—worlds apart—from man, as we recollect the animal development which must follow. These creatures, which play about like human children while they are young, sink afterwards into animality; the last gleams of human resemblance disappear in them, as we noticed in the chimpanzee obliged to change into a beast.

CONCLUSION

The countenance, the voice, and the play of animals, all these move the mind of man deeply, when he considers human destiny and the destiny of the animals. Man has to feel what a marvellous favour has been granted to him, in the universe, in that he can stand there open and sensitive, still incomplete, but capable of growth, still germlike, but with the prospect of maturity; as yet an imperfect reflection of the eternal entelechy, but endowed with the power of striving towards its ever clearer manifestation.

This is the badge of the nobility of man: that his bodily form can become continually more transparent, that the spiritual countenance which is dwelling there behind the course of his destiny looks out with increasing clearness through the face.

This is the world's gift to man; his entelechy can imprint itself in deeds, in thought, in every step, every word and every tone of the voice, in each event of his life, on each feature and gesture.

To become human means to set the seal of the spirit of the Ego upon all the earthly human life—that " I " which of its own free choice—and out of its wider survey—interweaves itself in the destiny of the earth. To become human means to bear the destiny of the earth onwards to the future stages of its existence. To be man is to know the animals and all the creatures on the earth; it is to recognise man's responsibility towards these beings, once of the same order as himself, but now obliged to live beside him in an incompleteness which never ceases calling to man—warning him to make himself worthy of the trust reposed in him.

He who has learnt to understand the silent speech of human life will sooner or later become aware in Rudolf Steiner's works, and in the course of his life, of something which has been taking place beside him unnoticed. He will look into this mirror, and before the boundless earnestness of the call which meets him, he must needs be silent. He will know what is lacking to him still, and how infinitely far behind he stands compared with the greatness which went by before his eyes; but he will have been allowed nevertheless to see what *Man* is, and to witness an immortal example of his life on earth.

NOTES

which also serve as a key to books consulted.

1 (to page 5).—Usually Klaatsch is quoted as the first to have spoken of the primitive nature of the human hand. Compare his posthumous work, *Werdegang der Menschheit* (Evolution of Mankind) and Schwalbe's attack on him in *Kultur der Gegenwart* (Modern Culture) III, 5, *Anthropologie*, 1924, pp. 271 and 311. The idea, however, occurs as early as 1887 in Karl Snell's posthumously published lectures on the descent of man (pp. 131 onwards). Read also Alsberg (1902, pp. 45 and 46) and Knauer (1914, pp. 8–9). Quite recently Dacqué (1924, pp. 338–41) has come forward as a supporter of Klaatsch.

2 (to page 10).—Important material relating to the persistence of fœtal conditions in the human body may be found collected in Bolk's book (1926). This author concludes his work with the sentence, "To a certain extent we represent the suckling forms of our progenitors."

3 (to page 13).—Friedenthal's *Sonderformen der menschlichen Leibesbildung* (Forms peculiar to the human physical organisation) also contains an abundance of the most valuable data concerning " retrograde " forms in man. In the 1917 edition (p. 53), he says: " For the human race, the retention of the characteristics of youth in the course of the history of the species is the rule, whereas, most of the other animals show a constantly wider departure from their youthful forms along the lines of a one-sided differentiation. Whenever in the human race the way to a one-sided differentiation is taken, we may be certain that it turns out sooner or later to be an error; it is adjusted by the dying out of this variation in human form. In this sense, therefore, among human beings, the more youthful form is also the form dominant for the future...."

4 (to page 14).—Details as to the phenomenon of arrested development in the human body occur in the following authors:

Skull	R. Virchow (1870), J. Chr. G. Lucae (1873), Eimer (1897, III, pp. 113–20), Kollman (1905).
Dentition	Klaatsch (1920, pp. 27–31), Naef (1926, pp. 96–7), Selenka (1898).
Brain	Bolk (1927).

Nose	Knauer (1914, p. 34), Schwalbe (1924, p. 231).
Larynx	Friedenthal (1910, 1926).
Skin	Meirowsky (1920).
Muscles	Klaatsch (1900), Knauer (1914).
The Axes of the Body	Bolk (1927, pp. 28-34).
Internal Organs	Westenhoefer (1926).
Embryonal Organs	Stratz (1906).

5 (to page 15).—The method of cognition used by Rudolf Steiner in his investigations has repeatedly been described by him in detail, and with the support of an expression of Goethe's has, on occasion, been characterised as "sensible-supersensible vision" ("sinnlich-übersinnliches Schauen"). It is quite distinct from any kind of dreamy perception through the aid of a medium, or anything of the sort, since it combines, not with dulling, but with heightening of consciousness. The three stages through which this method of cognition advances, Imagination, Inspiration and Intuition, are first described in the journal *Lucifer Gnosis* (1905, Numbers 32-4) and during the two following decades are often described again, finally in the book published in collaboration with Dr. I. Wegman (1925). See List of Authors quoted.

6 (to page 16).—The world of "Universalkräfte" (cosmic forces), streaming in from the periphery, in its polarity to the "Zentralkräfte" (central forces), as known to physics and radiating from the earth, was first described in greater detail by Dr. Steiner in a course on Optics, 1919. It corresponds to what is called in the other writings the world of the "Bildekräfte" (formative forces), in which man shares through his "etheric body." See the author's pamphlet *Der Bildekräfteleib der Lebewesen als Gegenstand wissenschaftlicher Erfahrung* (The Formative-forces-body of the Living Creature as the subject of Scientific Experience) (1924, p. 32 *et seq.*) and the essays in the weekly *Goetheanum* (III, pp. 172 and 236)—The description of the being of man—not only of his body—from the point of view of his threefold nature, was first given by Rudolf Steiner in his book *Von Seelenrätseln* (Riddles of the Soul) (1917, p. 230). There we read: "I may say, that I am here recording the results of an investigation both spiritual and scientific continued throughout thirty years." A short exposition by E. Kolisko of the importance of this fundamental description occurs in the monthly *Die Drei*, 1st year, 1921, pp. 541 *et seq.*

7 (to page 17).—Dr. Steiner has again and again characterised the perpendicular as the essentially human, the horizontal as the essentially animal position. He has then brought these two directions into cosmic relation (Sun, Moon and Earth), and has emphasised the great importance of the circumstance that man in sleep takes the horizontal position which is characteristic of animals (Astronomical Course, 1919, Lecture VII, onwards). Dr. Guenther

Wachsmuth (1924, pp. 77-83; 1926, pp. 82-91) has brought these two main directions into connection with the etheric currents of the earth. Here Rudolf Steiner opened an entirely new field of scientific investigation. For the purpose of the present book it has been necessary to concentrate on the more purely physiognomic of these aspects which are imaged in the pillar and the dome.

8 (to page 18).—Birds hold a peculiar position, inasmuch as their organisation connects them, not with gravity, but with the air. In this respect they contrast with the mammals, especially the quadrupeds. According to Rudolf Steiner's description (1921, pp. 78 *et seq.*; Ilkley Lectures, 1923, English edition, *Education and Modern Spiritual Life*, Ch. 8. London 1954; and elsewhere), the quadrupeds are to be understood as a specialised development of the human limb-and-digestive system, and the birds as metamorphoses of the human head (compare E. Kolisko, 1926, pp. 255 and onwards). Rudolf Steiner frequently stated that the birds have in many respects anticipated evolution, and that they therefore display a picture of future form conditions (e.g., Cycle 13, Lecture I, p. 9). In spite of this, however, there is justification for the description of birds which was given earlier; the forms taken by their limbs can be described as a specialised deviation from the basic type-form, the human hand. They are ranged with the mammals in contrast with man; for the bird's wing too has become a tool.

9 (to page 19).—In the Anthropoids, especially in the young, the eyes are so placed as to afford a free vision in front. In other animals, too, for example in the Tarsius (Gespenstmaki) and in the owls, there is a similar direction of sight. However, in none of these cases does the crossing of the axes of sight in the object—the decisive factor for human sight—appear to occur.

10 (to page 23).—An excellent older treatise on the difference between the extremities of man and of ape is that by Lucae (1865). The same author then says (1873, p. 15): " After examining a number of apes of the old and new worlds I have proved in detail that the bones of the hind and fore extremities do not justify Huxley's hypothesis " (i.e. the hypothesis that the differences between man and anthropoid ape are slighter than those between the latter and other apes.—H.P.), " as *all* apes have at the hind extremity a fully developed *thumb* which can be used in opposition to the other fingers; it is a perfect organ for grasping, a prehensile foot; while in place of this man has only a great *toe*, and therefore a *supporting foot*. . . . Man's foot seems just as ill- and just as well-suited for seizing and holding as a forearm pressed against the chest." The foot of man is an entirely self-reliant construction. Weidenreich's detailed monograph (1921) proves this. In coming to the conclusion that the ancestor of the Anthropoids possessed " right from the outset long lower extremities," when it exchanged the climbing for the walking method of life, the author admits, perhaps involuntarily, what is maintained above—viz., that no animal foot can have been the prototype of the human foot.

11 (to page 24).—Very valuable material relating to this will be found in

Knauer's work, 1914. The following sentences occur, illustrating the metamorphoses effected by the forces of gravity on the lower human organism. "In the new-born babe and especially in the fœtus the blades of the thigh-bones are turned sharply upwards as in the apes, and incline backwards more and more as the child learns to walk. At the time of puberty, however, we can recognise, with the commencement of strengthening of the musculature of the abdomen and generally of the whole body, that the thigh-bones again tend to take a somewhat upward direction, they become more steeply inclined" (p. 47). "The heavy pressure of the body from above on the base of the *os sacrum* has led to the perfecting of the promontory. Even among the Anthropoids it is scarcely permissible to speak of the formation of a true promontory. In the new-born human being it is not present either" (p. 50). "A development peculiar to man, connected with the excessive contraction of the muscles (of the lower thigh), and consequently with the acquisition of the upright gait, is the appearance of the Achilles tendon, produced by the merging of the end sinews of the soleus and the gastrocnemicus. The merging never occurs in apes" (p. 133).

12 (to page 36).—Particularly instructive is the history of the celebrated skull from Le Moustier in Dordogne. It was excavated in 1908 by Hauser and Klaatsch, and immediately pieced together as well as might be on the actual site of the find. It had a most unfamiliar appearance; from huge eye-cavities a beast-like face with an excessively protruding lower jaw peered out at the beholder. Later on in Berlin the skull had to be taken to pieces and put together again more carefully (Krause)—this time it was much more human in appearance. Quite recently, however, it was taken to pieces once more, and reassembled with infinite care by Weinert; a skull now appeared like that of other "primitive men," without a trace of pithecoid characteristics (cf. Weinert's monograph, 1925).

13 (to page 46).—As early as 1863, and again in his posthumously published lectures (1878), Karl Snell used an extraordinarily striking comparison: "If an ancestral lord could review what exists of his posterity after many generations, he would find among them, besides a few lords, many others living in more modest, or even in quite reduced circumstances. Man is the lord of creation. If he follows his genealogical tree backwards, he finds none but creatures, in whom the capacity for a progressive development—right on to the more inward universality of a race endowed with reason—was always maintained unbroken, and all of whom consequently belonged to the root stock. If, however, one of these far earlier generations of the root race could review its now living posterity, it would find among them, not only men, but many creatures, passing their lives in all stages of bestial limitation" (pp. 113-4). Read about Snell in H. H. Frei's excellent essay, 1922.

14 (to page 50).—The fundamental differentiation of the human being into members was first set out in the book *Theosophie* (see list of Authors), 1904, also

frequently later, and always from new points of view. Mention should be made here especially of the account given in *Grundlegendes für eine Erweiterung der Heilkunst nach geisteswissenschaftlichen Erkenntnissen. Fundamentals of Therapy* (1925).

15 (to page 51).—The genesis of the earth has been described from the viewpoint of the aggregate conditions and of the various ethers, in Dr. Wachsmuth's work on the Etheric Formative Forces: which is of fundamental importance for the new anthroposophical natural science (1924; second edition, 1926, pp. 41-56) and in the second volume (1927, pp. 26-36). See also his *Erde und Mensch* 2nd ed. 1952

16 (to page 51).—This table is taken from Rudolf Steiner's lecture cycle "The Secrets of the Biblical story of Creation," given at Munich, 1910, to members of the Anthroposophical Society. Translated into English: *GENESIS. Secrets of the Bible Story of Creation*. London, 1959.

17 (to page 57).—The evolution of man and animal has been described by Rudolf Steiner from many and various points of view. Particularly surprising and impressive is the description given in connection with the threefold character of the human organism which Albert Steffen has summarised in Dr. Steiner's *Lectures to Teachers*, Christmas, 1921 (English edition London, 1948, Chapter 5). " The head with its dome-like skull, with its grey brain matter, only slightly differentiated from cellular ganglia, and with its fibrous white matter deeper within—can be likened to the lowest grades of the animal kingdom, to the shell-fish. The chest organism, which comes principally under the dominion of the spine and the lymphatic glands, can be likened to animals of the middle grades, to the fish. The metabolic organism to the animals of the highest stage (for instance cow or camel). So that we get a threefold division of the animal kingdom in reference to the human organism. . . . Human evolution originally arose out of something which became later the head. It is the head of man which has taken the longest to evolve, and embryology bears this out. Actually the organisation of the head arose at a stage of life now represented by the shell-fish (oysters, etc.). These now represent in completely changed conditions of the earth that which man was in primitive epochs. In the course of his evolution he has thrown these organisms out of himself and developed further. They have to a certain extent remained behind.

" The fish species exhibit a second stage of man's development. The organisation of the fish began later than that of man. By the time it arose man could already draw forth from his own rhythmic organism those impulses which the fish had to draw from its environment. Thus the organisation of the intermediate animals was added afterwards to that of man, who had already reached a definite stage.

" And lastly, when man had already evolved his limb and metabolic system, the higher animals arose.

" The current, and still general, view of the descent of man is only valid for man in so far as it concerns the head. For this does take its root in forefathers, who in their physical constitution, not their spiritual, bore a remote resemblance to the lower animals of to-day, but differ again from them in that these are evolving under absolutely different circumstances.

"The intermediate and higher animals are of more recent origin than man, and to look upon them as his progenitors is entirely unjustifiable."

That such a conception is essential for a new Animal Morphology, Dr. Kolisko (1926, p. 259) has shown with great beauty and clarity. In the present book this must regretfully be left aside, so as not to confuse the representation of the four periods of the earth: to keep this in sight a firm stand must be taken.

18 (to page 74).—The astounding size of the brain in all the embryos of amniotes has always been the crucial problem for present-day ideas of descent. To get round it it was said that the brain in the embryo must needs be in advance of the other organs: its development—being an extraordinarily lengthy process—has to be started beforehand, as it were, and on that account, even in the early stages of the germ growth, it appears disproportionately great in size. But even if certain anachronisms (Heterochronien) in germ history actually occur (Meinert, Keibel), this must not be relied on too much. It would be attributing human methods to Nature to say that it is obliged to get going early with difficult work. Where necessary Nature can adapt its extraordinary powers, as in the creation of thousands of eggs in the ovary within a few days. No, the mighty development of the head, in the earliest embryo stage indicates unmistakably the true ancestral form whose head, still "unearthly", was yet wholly in the likeness of the cosmic environment. The head retains this its cosmic signature for the longest time, and indeed in other respects too it retains the past.

19 (to page 93).—Very pertinently Buytendijk, the Dutch psychologist, writes: "without doubt instincts are hereditary habits based upon inherited perception and function-capacity, nevertheless, it is incorrect to suppose that in the course of the individual's life experience these habits cannot be modified. Wherever the life functions are such that they offer a variety of perceptions and of behaviour, a plasticity of instinct also appears which is in correspondence with them. Nevertheless, the capacity for variation of instinct is wholly determined through instinct itself. The laws also are inherited according to which habits of animals—repeated or changed under certain conditions—adapt themselves" (p. 76 of André's German edition). "The experience possible to an animal is wholly built up within instinct: it belongs characteristically to its instinct" (p. 37).

20 (to page 96).—A series of lectures, "Die Offenbarungen des Karma" was given in Hamburg, 1910, by Rudolf Steiner to members of the Anthroposophical Society. In the second lecture he said: . . . "The peculiar phenomenon comes to light, that originally man and animal were similarly endowed; and if we were to go back to the Old Saturn evolution, we should find that there was

absolutely no difference between human and animal development. The two were then absolutely alike. What then happened in the meantime, that the animal now brings with it into existence all sorts of capacities while man is really a clumsy being when he comes into the world? ... Man has not wasted these powers, which at the present time the animal manifests as external capacities; he has only transformed them, but into something which differs from what the animals possess. The latter express them in external capacities; beavers build their homes and wasps their nests, etc. Man has transformed and incorporated within himself the same forces which the animals manifest in this way, and by this means he has brought into being what we call his higher human organisation. In order that man should be able to walk upright, in order that he should have a more perfect brain and, generally speaking, in order that he should have a more perfect inner organisation, certain forces were necessary, and these are the same forces with which the beaver constructs his dwelling.... We have our beaver building within us, and therefore we are no longer able to manifest these forces outwardly in the same way." English translation: *Manifestations of Karma*, London, 1947.

21 (to page 104).—Out of the many descriptions Rudolf Steiner gave, the one quoted here is from a public lecture at Arnhem, Holland, July 21st, 1924, entitled: "Was kann die Heilkunst durch eine geisteswissenschaftliche Betrachtung gewinnen?" English translation: *Spiritual Science and the Art of Healing*. London, 1950, p. 33 et seq. "Man has ... an external physical organisation which is perceptible by means of the outer senses, and whose manifestations can be comprehended by the reason. Besides this physical body there is also the first supersensible body of the human being: *the etheric or life-body*. These two principles of the constitution of man serve to *build up* (integrate) the human organisation. They physical body is continually renewed as it casts off its substance. The etheric body which contains the forces of growth and of assimilation—is, in the entirety of its constitution, something of which we can gain a conception when we behold the growing and blossoming plant-kingdom in the Spring; for the plants as well as human beings have an etheric or life-body. In these two members of the human organisation we have a progressively constructive development.

"In so far as man is a sentient being, he bears within himself the next member, *the astral body* (we need not feel that such terms are objectionable; we should perceive what they reveal to us). The astral body is essentially the mediator of sensation, the bearer of the inner life of feeling. The astral body contains not only the upbuilding forces, but also the forces of *destruction*. Just as the etheric body (or call it what we will, but there it *is*), makes the being of man bud and sprout, as it were, so all these processes of budding are continually being disintegrated again by the astral body; and just because of this, just because the physical and etheric bodies are continually being *disintegrated*, there exists in the human organisation an activity of *soul and Spirit*. ... But the nerve process is in a continual, though slow, state of dissolution; and because it is so, because the physical is always being dissolved, a place is set free for the Spirit and soul.

"In a still higher degree is this the case as regards the actual Ego-organisation, by means of which man is raised above all the other beings of Nature surrounding him on the Earth. The Ego-organisation is essentially bound up with katabolism: it is of greatest moment in those parts of a human being that are in a state of *disintegration*."

22 (to page 111).—Wasmann (1900, p. 47) gives a definition of animal experience which in its way is exact; he says of ants: "Only a faculty for sentient knowledge and response which—being influenced by outer sense perceptions, and by inner subjective states of feeling—is the cause of most variable and arbitrary behaviour—such a faculty alone can provide a satisfactory psychological solution."

But his expression "sentient" faculty of knowledge first comes to have meaning if the evident *picture* experiences of an animal are spoken of in the way Rudolf Steiner did.

23 (to page 115).—The difference between knowing and understanding was first noticed by G. Carus (1866, p. 91). He says: "even the young child's way of learning to know the world he lives in is quite different to an animal's. Of the latter it can never be said characteristically: it 'learns' to know, but much rather that it knows directly of that in the world which concerns its own organisation. And that is why it never achieves a true understanding, able to fit a sense impression with an idea."

In the same way the writer in the preceding third chapter has used the expression "behaviour" of an animal to contrast with the "deed" of man.

24 (to page 120).—This connection is brought out in a beautiful way by Oswald Spengler (1920, pp. 422–3, 451 onwards), with convincing examples of it.

25 (to page 122).—On the basis of his experiments with spiders, Hans Volkelt (1914) acutely remarks: "It is not the separate parts and details of the world surrounding it that direct the creature's original activity (in response); but rather that its response is governed by the *totality* of the impressions" (p. 63). Volkelt was forced to suggest that animal consciousness is not composed of shapeless atom-like contents (as in man) but is built up by a peculiar synthesis quite alien to man. "Every moment a wide range of physical impressions—perhaps all that is present at that moment—is contracted into one *complex of qualities*. This complex contains all details in itself, and yet no detail is separated and defined, but all are con-fused in the specific complex" (pp. 89–90). If we try to grasp the idea of such a complex, "alien to man," which reproduces an entire situation we have hardly other choice than the pictorial reproduction, symbolic yet specific, which is characteristic of the *Moon-consciousness* as described by Rudolf Steiner (see previous pp. 101-2).

26 (to page 130).—If this is not taken thus concretely there is no escape from anthropomorphisms, which veil the essential in animal psychology. B. Wasmann (1900, p. 82) writes: " Nowhere is there a mechanical agreement as to plan but every ant clearly decides for itself (!) in each case, of its own will to build, and its own plan of building(!)." " The most industrious and skilled worker ant has also usually the most imitators; her zeal, as it were, urges the others forward and leads their building energy in the same direction." The question is how does it come about that her industry appeals to the others, and that the others' desire to build is guided in the same direction: it is unthinkable without a motivating group-Ego which is common to the individuals concerned.

27 (to page 133).—The penetration of the animal's perceptions with bodily feeling has often been described by Rudolf Steiner. In his lecture at Arnhem (July 21st, 1924), referred to in Note 21, he said (p. 44), " The sense-life of man is entirely different from that of the animal. When the animal perceives something with its eyes—and this can be shown by a closer study of the structure of the eye—something takes place in the animal which, so to say, goes through the whole of its body. It does not happen like that in man. In man, sense-perception remains far more at the periphery, is concentrated far more on the surface. You can understand from this that there are delicate organisations present in animals which, in the case of the higher species, are only to be found in etheric form. But in certain of the lower animals you find, for instance, the xiphoid process which is also present in higher animals, but in their case it is etheric; or you may find the pecten or choroid process in the eye. The way in which these organs are permeated by the blood shows that the eye shares in the whole organisation of the animal and is the mediator to it of a life in the circumference of its environment. Man, on the other hand, is connected with his system of nerves and senses quite differently and therefore lives, in a far higher sense than the animal, in his *outer* world, whereas the animal lives more *within itself*."

28 (to page 148).—In the *Philosophy of Spiritual Activity* (English translation, 2nd Ed., 1922, p. 95) we read: " The perception of the table has produced a modification in me which persists like myself. I preserve the capacity to reproduce an image of the table. ... Modern psychology terms this image a ' memory-idea.' Now this is the only thing which has any right to be called the idea of the table."

Here, as elsewhere, Rudolf Steiner combats the error that a forgotten experience is retained as an image somewhere in the subconscious. Only the ability to reproduce the image is retained. This is what memory consists of. It is a creative deed (cf. *Von Seelenrätseln* (Riddles of the Soul), 1917, pp. 199-201).

The animal does not "remind" itself; it is reminded. A sense impression evokes an experience that is similar to an earlier one, and this is not a picture of the earlier experience reproduced by effort (cf. *Outline of Occult Science*, 1922, p. 27).

The essential quality of human memory is that it calls to mind. That this is truly an activity of the Ego is shown more specially in a state of fatigue, when one tries in vain to recollect something. Then perhaps later, due to some outer occasion, one can recall the forgotten matter. The animal can be imagined in a similar case: for it is dependent on the reawakening of the experience through the outer sense impressions. To be reminded is not to recall or recollect. To be able to recollect the Ego is needed; " presence of mind is needed. . . ."

LIST OF AUTHORS REFERRED TO IN THE TEXT

Hans André, Der Wesensunterschied von Pflanze, Tier und Mensch ..	1924
Moritz Alsberg, Die Abstammung des Menschen und die Bedingungen seiner Entwicklung	1902
K. E. v. Baer, Über Entwickelungsgeschichte der Thiere. Erster Theil	1828
Reden und kleinere Aufsätze	1876
M. Boule, L'homme fossile de La Chapelle-aux-Saints. Annales de Paléontologie, Band VI und VII	1911–1913
L. Bolk, Das Problem der Menschwerdung	1926
F. J. J. Buytendijk, Die Weisheit der Ameisen. Deutsch von H. André	1925
K. G. Carus, Vergleichende Psychologie	1866
E. Dacqué, Urwelt, Sage und Menschheit	1924
Th. Eimer, Die Entstehung der Arten. III Band	1897
Elbelt zit. nach O. Hamann s. d.	
H. H. Frei, Die Schöpfung des Menschen von Karl Snell. Monatsschrift " Die Drei " 2. Jahrgang Heft 12	1922
H. Friedenthal, Beiträge zur Naturgeschichte der Menschen. 5. Lieferung: Sonderformen der menschlichen Leibesbildung	1910
Die Sonderstellung des Menschen in der Natur	1926
J. W. v. Goethe, Physiognomische Fragmente. Naturwissenschaftliche Schriften, herausgegeben von Rudolf Steiner. 2 Band	1887
W. Giesbrecht, Crustacea. In : Handwörterbuch der Naturwissenschaften. II Band	1912
W. K. Gregory, The origin and evolution of human dentition. Journal of Dental Research. Vol. II	1920
E. Haeckel, Generelle Morphologie der Organismen	1866
F. Keibel, Das biogenetische Grundgesetz und die Cenogenese. Merkel und Bonnets Ergebn. d. Anat. u. Entw., Band 7	1897
H. Klaatsch, Der kurze Kopf des Musculus biceps femoris. Sitzungsber. d. Königl. Preuss. Akad. d. Wiss.	1900
Der Werdegang der Menschheit, n. d. Tode herausgegeben von A. Heilborn	1920

S. Knauer, Ursachen und Folgen des aufrechten Ganges des Menschen 1914
W. Köhler, Intelligenzprüfungen an Anthropoiden I 1917
G. H. R. v. Königswald, Begegnungen mit dem Vormenschen. Köln.. 1955
E. Kolisko, Die Dreigliederung des menschlichen Organismus. Monatsschr. "Die Drei," I Jahrg. 1921
Gedanken zur anthroposophischen Tierkunde, in: "Gäa Sophia," Jahrb. d. naturwiss. Sektion d. Freien Hochschule f. Geisteswissenschaft am Goetheanum, Band I 1926
J. Kollmann, Neue Gedanken über das alte Problem von der Abstammung des Menschen. Corresp.-Blatt d. deutsch. Ges. f. Anth., Ethn., Urgesch. 36 Jahrg. 1905
K. König, Einige geisteswissenschaftliche Betrachtungen über die Eihüllen und die erste Anlage des Menschenkeimes. Monatsschr. "Natura" I 1927
Versuch einer Darstellung der jüngsten menschlichen Embryonalentwicklung. "Gäa Sophia," Jahrb., II Band 1927
E. Kretschmer, Hysterie 1923
J. Chr. Lucae, Hand und Fuss. Abh. d. Senckenb. Naturf. Ges... .. 1865
F. Meckel, Entwurf einer Darstellung der zwischen dem Embryonalzustande der höheren Tiere und dem permanenten der niederen stattfindenden Parallele. Beitr. z. vergl. Anatomie, Bd. II 1818
E. Mehnert, Die individuelle Variation der Wirbeltierembryonen. Morphol. Arb., Bd. 5 1898
E. Meirowsky, Die angeborenen Muttermäler und die Färbung der menschlichen Haut im Lichte der Abstammungslehre. Naturw. Wochenschr., N.F., Bd. 19 1920
Fritz Müller, Für Darwin 1864
A. Naef, Über die Urformen der Anthropomorphen und die Stammesgeschichte des Menschenschädels. "Die Naturwissenschaften," Wochenschr., 14 Jahrg. 1926
H. Petersen, Über die Bedeutung der aufrechten Körperhaltung für die Eigenart des menschlichen Umweltbildes. Ebenda, 12 Jahrg. .. 1924
E. Pfeiffer, Die geologische Erdentstehung im Lichte der Geisteswissenschaft. Jahrbuch "Gäa Sophia," Bd. I 1926
H. Poppelbaum, *Tier-Wesenkunde* (1938) 2. Auflage 1954 Dornach. *Begriff und Wirkungsweise des Aetherleibs*. Anthroposophisch-medizinisches Jahrbuch III. Dornach 1952
A. Portmann, *Das Tier als soziales Wesen*, Zürich 1953
L. Rütimeyer, Gesammelte kleine Schriften allgemeinen Inhalts .. 1898
A. Schopenhauer, Über den Willen in der Natur 1835
G. Schwalbe, Die Abstammung des Menschen und die ältesten Menschenformen. In: Kultur der Gegenwart. Band "Anthropologie" 1923
E. Selenka, Menschenaffen (Anthropomorphae) 1898

Karl Snell, Die Schöpfung des Menschen 1863
Vorlesungen über die Abstammung des Menschen. Aus dem Nachlasse herausgeg. von R. Seydel 1887
O. Spengler, Der Untergang des Abendlandes, Bd. I 1917
Rudolf Steiner, Grundlinien einer Erkenntnistheorie der Goetheschen Weltanschauung 1886
(*The Theory of Knowledge Implicit in Goethe's World Conception. Fundamental Outlines with Special Reference to Schiller*, New York, 1940)
Die Philosophie der Freiheit 1894
(*Philosophy of Spiritual Activity. Fundamentals of a Modern World Conception*. London, 1921, and later editions)
Goethes Naturwissenschaftliche Schriften (*Goethe the Scientist*. New York, 1950)
Theosophie. Einführung in übersinnliche Welterkenntnis und Menschenbestimmung 1904
(*Theosophy. An Introduction to the Supersensible Knowledge of the World and the Destination of Man*. London, 1954)
Die Theosophie des Rosenkreuzers (*The Theosophy of the Rosicrucians*. London, 1953) 1907
Lucifer Gnosis, Zeitschr., herausgeg. von Rudolf Steiner .. 1904–1908
Aus der Akasha-Chronik (*Cosmic Memory. Pre-History of the Earth and Man*. New York, 1959) 1955
Blut ist ein ganz besonderer Saft 1907
(*The Occult Significance of Blood*. London, 1926) .. ..
Die Geheimwissenschaft im Umriss 1910
(*An Outline of Occult Science*. New York, 1923, and later editions. P. Nos. quoted are from the 1923 ed.)
Die Offenbarungen des Karma (*Manifestations of Karma*. London, 1947) 1910
Unsere atlantischen Vorfahren 1909
(in *Atlantis and Lemuria*. London, 1911, and *Cosmic Memory* as above)
Wege zu einem Baustil (*Ways to a New Style in Architecture*. London, 1927) 1914
Von Seelenrätseln 1917
Das Karma des Materialismus 1916
Das Verhältnis der verschiedenen naturwissen-schaftlichen Gebiete zur Astronomie 1919
(Herausgegeben von der astronomischen Sektion der Freien Hochschule f. Geistesw. in Dornach, 1926.)
Der Lehrerkurs Rudolf Steiners am Goetheanum, bearbeitet v. Albert Steffen 1921
(*Lectures to Teachers*. London, 1948)

Esoterische Betrachtungen karmischer Zusammenhänge I-IV (*Karmic Relationships. Esoteric Studies.* London, I-IV. 1955-7) 1924
Was kann die Heilkunst durch eine geisteswissenschaftliche Betrachtung gewinnen? (*Spiritual Science and the Art of Healing.* London, 1950) 1924
Mensch als Zusammenklang des schaffenden, bildenden und gestaltenden Weltenwortes. (*Man as Symphony of the Creative Word.* London, 1945) 1923
Initiations-Erkenntnis (*Evolution of the World and Man.* London, 1926) 1923
u. I. Wegmann, Grundlegendes für eine Erweiterung der Heilkunst nach geisteswissenschaftlichen Erkenntnissen 1925
(*Fundamentals of Therapy. An extension of the Art of Healing through Spiritual Knowledge.* London, 1925, New York, 1938)
O. H. Schindewolf, Das Problem der Menschwerdung. Jahrb. der Preuss. Geologischen Landesans. Bd. 49 1928
C. H. Stratz, Zur Abstammung des Menschen 1906
R. Virchow, Menschen- und Affenschädel. Samml. wissensch. Vorträge 1870
Hans Volkelt, Über die Vorstellungen der Tiere. Arb. z. Entw.-Psychologie, Bd. I, 2 1914
G. Wachsmuth, Die ätherischen Bildekräfte in Kosmos, Erde und Mensch. I Bd. 1924
(*Etheric Formative Forces in Cosmos, Earth and Man.* London, 1932)
E. Wasmann, Vergleichende Studien über das Seelenleben der Ameisen und der höheren Tiere 1900
F. Weidenreich, Der Menschenfuss. Zeitschr. f. Morphol. u. Anthropol. Bd. 22 1921
Apes, Giants and Man. Chicago, 1946
H. Weinert, Der Schädel des eiszeitlichen Menschen von Le Moustier in neuer Zusammensetzung 1925
M. Westenhöfer, Der Mensch—das älteste Säugetier. Mitteilungen der Anthropolog. Gessells, Wien. Bd. 5 1927